The Behavior of Domestic Animals

The Behavior of Domestic Animals

BENJAMIN L. HART

University of California, Davis

W. H. FREEMAN AND COMPANY / NEW YORK

Library of Congress Cataloging in Publication Data

Hart, Benjamin L.
    The behavior of domestic animals.

    Includes bibliographies and index.
    1. Domestic animals—Behavior.   I. Title.
SF756.7.H37   1985   636   84-25893
ISBN 0-7167-1595-3

Printed in the United States of America

1 2 3 4 5 6 7 8 9 0   HL   3 2 1 0 8 9 8 7 6 5

This book is dedicated to my daughters
Rachel and Jessica
in appreciation of their love
for me and their animal friends.

CONTENTS

PREFACE

■

PURPOSES OF THIS BOOK

This book is a basic text dealing with the ethological, experiential, and physiological aspects of domestic animals. The material is presented with two purposes. One purpose is to offer a biological perspective on comparative animal behavior that uses familiar animals as resource material. Most of the readers of this book will have had a dog or cat as a pet; some will have owned horses or livestock as well. I feel that many students may prefer a study of animal behavior that uses animals with which they have interacted, rather than wild animals, as resource material. The fact that domestic animals represent a wide array of different mammalian types allows for a truly comparative approach.

The second purpose is to present behavioral information that is relevant to the care and management of both companion animals and livestock. The most intelligent, humane, and cost-efficient approach to the breeding, care, and use of domestic animals must be based in part on an understanding of their natural behavioral tendencies. Some animals are maintained primarily because of their behavior; this is true of cats, dogs, and horses. The more we know about their behavior, the happier we will be with them as pets, and the better off our pets will be with us. With the livestock that are kept for economic reasons, we can use behavior to our advantage or disadvantage. Studies have shown, for example, that behavioral stress may take an economic toll. Animals that display inadequate sexual performance or abnormal maternal care can also cause economic losses. The more we understand about all aspects of the behavior of animals, the more efficient and humane livestock production will be.

Is agriculture becoming too big and commercial to warrant concern over the behavior of livestock? Probably not. The North American and European style of animal agriculture with feedlot fattening of cattle and mechanized dairy operations requires more fossil fuels than when a few animals freely graze in a large pasture. Concern with the increasing costs of energy will change the way in which we raise animals. The most energy-efficient modes of obtaining animal food are hunting and herding, which characterized the existence of early humans at the time when domestication began. The trade-off is, of course, the amount of land used: the more old-fashioned system uses less energy but requires more land.

It is clear that animal agriculture is going to become less dependent on grain supplementation, which is available only at increasing energy costs for cultivation and fertilizers. As fuel becomes prohibitively expensive, we are likely to find it impossible to continue our intensive fuel-based agriculture. In addition, much of the world's land is too steep, too wet, too dry, or too cold to produce human food crops, but it can be efficiently harvested by ruminants. Production of livestock on this type of land is far below its potential.* As we drift back into a grass-only type of livestock production, there will be sacrifices in the production of animal protein, but energy economics will force some major shifts in this direction.** In order to efficiently raise livestock under more natural pasture conditions, it will be necessary

---

* Ward, G. M., T. M. Sutherland, and J. M. Sutherland. 1980. Animals as an energy source in third world agriculture. *Science* 208:570–574.
** Pimentel, A., P. A. Oltenacu, M. C. Nesheim, J. Krammel, M. S. Allen, and S. Chick. 1980. The potential for grass-fed livestock: Resource constraints. *Science* 207:843–848.

to utilize knowledge about their behavior. Through an integrated ethological, behavioral, and physiological approach, we can manage our countryside and animal friends better than we have in the past.

■

ORGANIZATION OF THIS BOOK

The book is organized in two sections, written in such a way that the reader can begin with either section. The first section (Chapters 1 through 5) describes the behavior that one observes in studying adult animals, including social, sexual, maternal, and feeding behavior. These species-typical behavioral patterns stem from the behavior that the ancestors of domestic animals acquired through centuries of natural selection. The second section (Chapters 6 through 9) outlines the present understanding of the theoretical background and determinants of behavior that regulate the expression of species-typical characteristics. These determinants include genetic manipulation, early experience, learning, and hormones. The determinants operate on behavior in general ways and can be considered somewhat apart from species membership. Much of the basic information comes from work on laboratory animals, especially rodents. However, I shall dwell primarily on those areas of research that are of importance in understanding domestic animal behavior. At the conclusion of each chapter dealing with determinants, there are some comments about the applied significance of the basic information presented. Current animal welfare issues are mentioned in a final chapter.

In my work, the animals' own behavior has posed the puzzle that has become the research question. Later, I have explored the theoretical background that related to the question. Thus, this book begins with the actual behavior of the animals and then presents the studies of the determinants and the theory concerning the behavior. While this sequence is most meaningful for some people, others prefer to examine the theoretical background before the behavior, and the book can easily be used either way.

The area of behavioral therapy of animals is specifically not covered in this book, even though a few clinical examples are woven into some of the practical application sections. The subject of behavioral therapy is treated elsewhere.*

■

ACKNOWLEDGMENTS

Many people have contributed facts, concepts, and insights about animal behavior to this book. First, I am indebted to my students for their intellectual contributions and for inspiring me to undertake the commitment to write it. Among others to whom I am indebted is my colleague, Dale F. Lott, who kept pointing out the value of looking at the behavior of domestic animals from the standpoint of evolutionary and adaptive concepts. Another colleague, Edward O. Price, patiently and painstakingly read every word of an early draft and uncovered numerous oversights, errors, and confusing sentences. This book has been immeasurably improved by the efforts of these two friends. My colleague and wife, Lynette A. Hart, spent endless hours on the manuscript and in discussions with me on various conceptual issues. My expression of gratitude to her is only a token representation of the major contribution she made to the completion of this book.

* Hart, B. L. 1978. *Feline behavior*. Santa Barbara, Calif.: Veterinary Practice Publishing Co.; Hart, B. L. 1980. *Canine behavior*. Santa Barbara, Calif.: Veterinary Practice Publishing Co.; Hart, B. L., and L. A. Hart. 1985. *Canine and feline behavioral therapy*. Philadelphia, Pa.: Lea & Febiger.

CHAPTER ONE ▪ INTRODUCTION

T he commonest mammals around us are domestic species. The animals serve the purpose of providing food, fiber, labor, recreation, personal safety, and companionship. These species originated from a very diverse assortment of wild ancestors, each of which was domesticated because of rather specific attributes: cattle and goats for milk production, sheep for fiber, horses for transportation, cats for rodent control, and dogs for companionship and assistance in hunting.

Domestication has been an active process occurring over hundreds of generations, during which man has controlled the breeding, care, organization of territory, and feeding of the target species. To increase milk production in cattle, for example, cows of breeds with high milk output were systematically favored as breeding stock. Over an extended period of time, animals have been selected that are manageable, thrive, and reproduce in artificial environments and possess a high quality or quantity of meat, wool, or other desirable attributes. By selective breeding of animals, human beings have accelerated and directed the genetic changes, or evolutionary shifts, within domestic species. The domestication process, then, in exerting biological change within a species far exceeds the effect of simply taming the behavior of individual animals.

We have tended to neglect the study of behavior in our domestic animals, partly because they are so common and therefore not as interesting as the more exotic species, and partly because they live in artificial environments and their behavior may be considered unnatural. The domestic animals in fact retain most of the innately determined behavioral patterns of their wild ancestors. This is quite evident when they are kept in more open, unrestricted areas that allow the display of more of their behavioral repertoire. The very factors that make domestic animals less interesting to study in some ways than wild animals create opportunities for learning something about animals that cannot be approached in the wild. The artificial environment gives us the opportunity to observe behavior much more closely than is possible in the wild. It also gives us the opportunity to analyze the physiological bases of behavior and to examine the effects of environmental manipulations. What we learn about animals in the domestic environment can then be extrapolated back to related wild species.

Domestic animals provide a diversified representation of mammalian species. We have animals whose life styles are centered around being either predator or prey, highly social or asocial, territorial or nonterritorial. Most species are polygamous in sexual orientation, but one important species, the dog, is basically monogamous. Maternal behavior among the various species ranges from that which is associated with strong maternal-infant bonds to that associated with weak bonds. Domestic animals even vary in their eliminative behavior, from species that are fastidious in their sanitary habits to those that are not the least bit selective in where they eliminate.

In the artificial environment our animals sometimes suffer in ways that are similar to what humans might experience. Some of our animals are subjected to the same abuses of early experience as we, resulting in antisocial behavior and psychosomatic disorders. We have found ways of controlling our animals through training, castration, and genetic manipulations. All these factors make the study of domestic animal behavior interesting and highly practical.

■

ORIGINS OF DOMESTIC ANIMALS

Why and How Animals Were Domesticated

To understand the behavior of domestic species it is useful to know something about the animals from which the domestic species were derived. With some animals, such as the dog and cat, we can look at existing wild species whose progenitors formed the ancestral stock and examine the natural context of various species-typical behavioral patterns. In other instances, such as with cattle and horses, the ancestral line has long since become extinct, and we must settle for the study of wild animals that are only related to the domestic species.

Domestic animals that have found their way back into reproducing in the wild, so-called feral animals, are also a source of information about the role and adaptive value of species-typical behavioral patterns. There are instances of contemporary feral-living groups of almost all domestic animals (McKnight, 1964), most notably horses, sheep, goats, dogs, and cats, and these animals are being studied with increasing interest.

Although the only domestic species discussed in this book at any length are dogs, cats, horses, pigs, cattle, sheep, and goats, many more species than these have been domesticated. Some others are reindeer, water buffalos, elephants, camels, donkeys, ferrets, rabbits, small laboratory animals, poultry, and honeybees.

Theories abound as to why various animals were initially domesticated (Isaac, 1970). One school of thought emphasizes economic and religious purposes. Early humans needed animals for food, clothing, and labor and for religious sacrifices. F. E. Zeuner, whose book (1963) on domestication stands as the most authoritative work in this area, maintains that Mesolithic humans would have found it far easier to obtain necessary supplies by hunting and trapping rather than embarking on experiments of taming unwilling animals that would reward their efforts only after several generations of selective breeding. Zeuner prefers the "biological" approach as an explanation of domestication. The daily behavior of early humans, he claims, made the appearance of domestication almost inevitable. The habit of keeping young animals, especially the ancestors of the dog, as pets was at the root of domestication. The scavenging habits of the wild progenitor of the dog brought it into contact with the human social medium, and from there it was a short step to humanity's befriending of the dog and using it as a hunting companion. Initial phases of domestication of other species probably did not occur until some form of agriculture was under way. Planted fields were highly attractive to herbivorous animals such as the ancestors of sheep, goats, and cattle, and this resulted in bringing these animals into more frequent contact with early humans.

The initiation of new hunting methods that accompanied the advent of crop planting may have been the trigger for one form of domestication. Downs (1960) noted that when hunting practices progressed from stalking and ambushing to driving wild animals into canyons or corrals made with sticks, brush, and posts, domestication became almost inevitable. With a successful hunt more animals would be corraled than could be eaten at once. If they were all killed, spoilage would be a problem. However, keeping live animals around would be a way to prevent spoilage, and the corral could come in handy as a holding pen. The cultivated crops of the hunters would have enabled them to feed the animals until they were needed for food. Archeological evidence reveals that these corrals were usually some distance from the villages, which meant that taking animals to the villages alive would also prevent spoilage. This in turn meant that the most easily handled animals would be saved the longest and the more aggressive and dangerous animals killed first. The first animals allowed to breed in the captivity of the corrals would, of course, be those most easily handled; hence we can see the origin of selective breeding for docility and reduced fear of humans.

Each of the various species that was domesticated possessed, even as wild animals, certain behavioral characteristics that made them desirable for maintenance under human care. Some species were preadapted for domestication (Hale, 1969). For example, herbivores that were highly social, nonterritorial and sexually promiscuous, such as cattle, sheep, and goats, were most easily maintained. These animals were also nonselective feeders and ecologically flexible, which meant they could be transported some distance or traded (Tennessen and Hudson, 1981). Animals with behavioral patterns that did not harmonize with early husbandry practices were not widely propagated. The archeological record shows that several ungulates, including moose and gazelle, were husbanded at one time, but the practice was later abandoned (Jarman, 1976).

The ancestors of dogs, which belonged to one of the species of wolf, were territorial, making them good watch animals. They were also highly social, which meant they had the capacity to take a subordinate role with humans, allowing them to dominate and control. Being adapted to pack hunting, the first wolf dogs were very useful as hunting companions. The wild ancestors of cats were territorial but had solitary life styles. Being solitary they hunted prey, such as small rodents, that they could kill alone without a pack. Their tendency to live on daily kills of small rodents made them especially valuable in controlling rodent populations around early humans' granaries.

The process of domestication has influenced both the morphology and the behavior of animals (Ucko and Dimbleby, 1969). Of course, the process is still continuing, with some recent changes occurring more rapidly than those in early human history. Zeuner divides the process of early domestication into five stages. In the first stage there were loose ties between animals and the social medium of humans. Interbreeding with the wild forms was common, and the animals retained a close similarity to the wild forms. In the second stage humans controlled the breeding and did not allow their animals to contact wild forms. The need to handle animals made it necessary to keep only the smallest ones and to select through breeding for reduction of body size and weapons. Behavioral characteristics such as docility, reduced fear of humans, and increased tolerance of confinement were selected as well. Later, in the third stage of domestication, humans became interested in increasing the size of their animals for greater meat production. They thus allowed some interbreeding with wild forms that were larger, but apparently they were careful to make sure that behavioral temperament did not revert back to that of the wild animals.

The fourth stage, which was an extension of the third, was a turning point. Presumably recognizing that they could enhance milk production, change fiber or wool characteristics, and alter body conformation by selective breeding, the early domesticators brought about some major bodily changes in their animals and eventually developed various breeds. Even before 300 B.C. there were distinctive breeds of domestic sheep, goats, and cattle in the Middle East. During this stage of domestication, when breeds were becoming standardized, interbreeding

with the wild ancestral species would have spoiled the characteristics so patiently cultivated through artificial selection. Thus
the wild forms would have been considered as enemies. This,
according to Zeuner, spelled doom for some of the wild ancestors, and they were systematically exterminated during what he
calls the fifth stage of domestication. As numbers of the wild
forms became extremely limited, they were absorbed into the
domesticated stock.

Zeuner does not mention this, but at the end of the fifth
stage of domestication, the domesticated species obviously still
retained all of the species-typical behavioral patterns related to
feeding, elimination, social interactions, and reproduction that
characterized the wild ancestors. Under farming and ranching
conditions that lasted until the turn of this century, animals
foraged for themselves, lived in social groups, and mated and
raised their own young more or less in the same fashion as the
wild ancestors would have. A sort of natural selection on the
farm maintained these essential aspects of behavior at peak performance. Animals that refused to mate or were poor mothers,
for example, did not reproduce themselves, whereas those with
early sexual maturity and a high rate of reproduction had the
most young.

Recently farming practices have changed so drastically that
it is appropriate to consider we are now in the sixth stage of
domestication. In this stage we are seeing the loss of the ability
of domestic animals to survive and reproduce themselves without human intervention. The species-typical behavioral patterns, so necessary for survival in the wild and on the old-fashioned farm, are no longer critical. Species-specific foraging
behavior is not necessary when we provide forage in cubes sup-
plemented with minerals. Social behavioral patterns necessary
for maintaining a rank-order dominance hierarchy are less important when animals are raised in pens or stalls. Courtship and
coordinated sexual interactions between males and females are
dispensable, even undesirable, when artificial insemination is
practiced. When the species-typical behavioral patterns that
were formerly necessary for individual survival and reproductive
success are no longer critical, then animals lacking the complete
genetic base for these behavioral patterns survive and reproduce
themselves quite readily. The result is the appearance of some
animals with weakened behavioral patterns related to forage selection, sexual behavior, and maternal responsiveness.

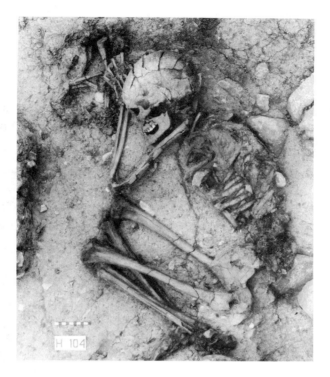

Figure 1-1 Excavation of a tomb in the Middle East (Israel) dating 10,000 to 12,000 years ago. A puppy skeleton is buried with a man. The hand of the human skeleton is lying on the thorax of the puppy, indicating that an affectionate bond existed between the puppy and the buried person. (From Davis and Valla, 1978.)

Wild Ancestors of Domestic Animals

Dog The dog was the first species domesticated. There are several archeological findings that reveal an association of wild canids with humans, beginning as long ago as 10,000 B.C. in North America, Europe, and Asia (Clutton-Brock, 1977).

The most recent documentation of the domestication of the dog is an excavation in northern Israel dating to about 12,000 years ago, showing a puppy skeleton in close association with a human skeleton. The human skeleton lay with its hand on the thorax of the puppy skeleton, indicating an affectionate rather than a gastronomic relation (Figure 1-1). There is no evidence for the domestication of other animals for at least one millennium later (Davis and Valla, 1978).

The most likely ancestor of the dog is the wolf, in particular the smallish Indian wolf (*Canis lupus pallipes*; Zeuner, 1963). Other canids, including dingos, jackals, and coyotes, will cross with the dog, however, and it is likely that today's dog possesses genetic inputs from several ancestral canid types. Most behaviorists use the wolf as the ethological reference point for many of the behavioral patterns of dogs. At this time the most is known behaviorally about the wolf of North America.

Goat   The goat appears to be the oldest domesticated ruminant, dating back to 8000 B.C. in the Middle East (Protsch and Berger, 1973). Of the possible wild ancestors, the bezoan goat (*Capra aegagus*) is the most important according to Zeuner. This ancestor still survives in small numbers in the Middle East.

Sheep   Before agriculture was fully developed, possibly as early as 7200 B.C., the sheep was domesticated with the aid of the dog; which served as sheep herder. The center of sheep domestication was in the Middle East, but old remains have also been found in southeastern Europe (Protsch and Berger, 1973). The first domesticated race stemmed from the urial group of ancestors, *Ovis orientalis*. This was one origin of the wool-type sheep. The moufflon group of ancestors from Asia formed another origin. The most primitive breed of sheep in existence today is the Soay breed, maintained in Scotland, which is closely related to the moufflon stock. The third wave of sheep to contribute to the domestic stock was of the Mesopotamian type. The Merino, known for its quality wool, belongs to this group. Neither the bighorn sheep of North America nor the Dall sheep of Alaska and British Columbia ever contributed to the origin of domestic sheep. The lowland urial sheep still exists in the Near East, and the moufflon type of sheep can be found today in isolated parts of western Europe and western Asia.

Cattle   The wild ancestor of domesticated cattle (*Bos primigenius*) ranged throughout Europe and the Near East (Isaac, 1962). The oldest dating of domesticated cattle is 7000 B.C., at a site in Greece. The ancestor of cattle, also referred to as the urus or aurochs, became extinct in the seventeenth century. The bison is one of the close wild relatives of cattle.

There are two species of domestic cattle, zebu or humped cattle, *Bos indicus*, and European cattle, *Bos taurus*. Whereas the European cattle are known to be descended from the wild aurochs, the origin of the zebu cattle is less clear. Zeuner claims that the zebu could well have descended from the wild Indian cattle, *Bos namadicus*, which is closely related to the primogenial breed. The domestication of cattle was a very important step in agricultural development because it is here, with all the trouble attendant to controlling such a large beast, that the economic advantages of domestication are evident.

Swine   Evidence of domesticated swine dates back to 7000 B.C. (Protsch and Berger, 1973). In Asia the local wild pig, *Sus vittatus*, was domesticated as a house pig where it could be confined and fed or allowed to find its food in the woods. The European wild pig, *Sus scrofa*, gave rise to the large herd pigs of Europe. In earlier times the two species were crossed, and modern breeds evolved from these two original sources. Wild forms of *S. scrofa*, known as the wild boar, and *S. vittatus*, known as the banded pig, still exist in isolated areas. The piglike African bushpig and the warthog have not contributed to domestic stock. The pig is recognized as a food animal, but it is interesting that it has been utilized for other purposes, such as treading seed into the ground after floods (Egypt and Greece), searching and grubbing for truffles (France), and as a hunting companion for pointing and retrieving.

Horse   Sometime between 4000 and 2500 B.C. the horse was domesticated from wild stock that lived in the Soviet Ukraine area. The horse was useful as a work animal and gave people great mobility, especially those who took up secondary nomadism combined with crop-raising. Later the horse-chariot combination became perfected and eventually ended up in Egypt. Much later, riding became popular. The wild stock that contributed most to equine domestication was the tarpan, which became extinct in 1851 in the Ukraine. The other wild race, the Przewalski horse, probably supplied some of the domestic breeds of central Asia and China. A few Przewalski horses survive today in zoological parks.

Cat   The cat is the most recently domesticated species of all mammals. This species is also the least altered morphologically and behaviorally from its wild ancestors. The true ancestor of the cat, *Felis libica*, is found throughout Africa. A closely related species, *F. silvestris*, the European wild cat, is still found in parts of Scotland. The domestic cat crosses with the European wild cat, but the latter is not believed to have been involved in the original domestication. Cats became properly domesticated sometime between 3000 and 2000 B.C. in Egypt, probably after they had been longtime intruders of human habitations in search of their usual prey, small rodents. Within a short period, cats became one of the most worshiped animals. Due to the embargo the Egyptians placed on these animals as sacred beasts, cats were late in reaching Europe in significant numbers. With the spread of Christianity cats were no longer considered sacred, and they became popular as household pets or companions. Later, in the fourth century A.D., the Roman ruler Palladius recommended they be used instead of polecats to kill moles that were damaging artichoke beds. About that time, according to Zeuner, they interbred with the local European wild cat.

Poultry   The red jungle fowl (*Gallus gallus*) is found in Asia and is the chief ancestor of the domesticated chicken. Anatomical evidence in the Indus valley suggests that full domestication had taken place by 2000 B.C., but it is not known whether the birds were initially used for sport or for food. The fowl served as a fertility symbol and also as a timekeeper, because it crowed reliably at the same time each morning.

The turkey (*Meleagris gallopavo*) that has become domesticated is from southwestern Northern America. It was already domesticaed when Cortez arrived, and it was introduced into Europe in 1523.

One can readily see some uses of animals in present society that are obviously different from the original ones. Although the cat was used earlier to control pests, it is now usually kept for companionship. The dog's current role includes some work tasks, such as pulling a sled, guiding the blind, or acting as a beast of burden, as well as companionship. Horses are used more for recreation now than for work. When large animals are used in recreational events such as rodeos, the relation with people sometimes seems to be based on the human participants subduing and dominating the animals (Lawrence, 1982). The persistent attraction of this theme in human society makes one wonder if it may have been a motivating factor in the domestication of such large animals as cattle and horses.

THREE PERSPECTIVES ON ANIMAL BEHAVIOR

There are three basic approaches to the study of animal behavior, the ethological, the experiential, and the physiological. It is useful to briefly examine how each of these perspectives relates in particular to our interest in the behavior of domestic species.

The Ethological Approach

The ethological approach represents the efforts of naturalists who have traditionally sought to understand behavior from the standpoint of its adaptive value for the animal in the wild. The innate behavioral patterns that are unique to each species, so-called species-typical behaviors, are of interest here. This area is particularly concerned with the survival value and the reproductive advantages gained by certain behavioral patterns and the genetic mechanisms that effect the behavioral patterns, that is, the evolution of behavior. Sociobiology is one of the current fields of interest stressing the way in which interactions between animals reflect the individual animal's attempt to maximize its genetic output. When we speak of a "cause" or "reason" for behavior displayed by an animal, we are often thinking of the evolutionary importance or the adaptive value of the behavior. This is referred to as the ultimate cause to differentiate it from a proximal cause that focuses on physiological or learning mechanisms.

In our discussion of domestic animals we shall be utilizing concepts from the ethological approach to examine a number of interesting species differences. Some of the behavior patterns that are largely under genetic control are very desirable on the domestic scene, whereas other patterns have proved quite troublesome. The use of threats and submissive gestures to settle conflicts, rather than outright fighting, is an example of a highly useful trait. The rejection of alien lambs by ewes is an example of a trait that was adaptive in the wild but poses special problems for the farmer who wishes to find a mother for an orphaned lamb.

Species-typical behavioral patterns of all domestic animals have undergone some genetic alteration through centuries of breeding. We have systematically and artificially selected as breeding stock animals that not only produce the most milk or wool but those that are inherently less fearful of humans, more docile to handle, and more tolerant of confinement. We see the weakening of some behavioral patterns because of lack of natural selection. Maternal behavior, for example is absolutely critical to the survival of an animal's offspring in the wild, but is not essential on the domestic scene because we step in. Since there is no particular selection for the genetic elements of maternal behavior, this behavior ends up being highly variable from animal to animal.

The Experiential Approach

We are concerned in the experiential approach with behavior acquired through early experience or learning rather than through natural or artificial selection. The emphasis is on similarities between species in the processes underlying acquired behavior. Research studies have employed the laboratory rat primarily, with the assumption that principles established for this species are applicable to other mammalian types. This assumption is inaccurate in some instances, but nonetheless, studies of learning in rats contribute to an understanding of domestic animal behavior.

Animals learn responses on the domestic scene that they never would have acquired in the wild (Kratzer, 1971). For example, dogs can be trained to attack and to withdraw from an attack on command. Horses are taught to carry us on their backs, and cattle line up for milking. Some interesting work in behavior now suggests that management of livestock species could be improved by the application of learning principles so an animal can make learned responses to show us its preference for lighting or housing temperature, perhaps saving on energy bills. The use of special techniques such as desensitization and counterconditioning (Chapter 8) have contributed to the growing field of behavioral therapy for problem behavior in dogs and cats (Hart, 1978, 1980).

The Physiological Approach

The physiological approach is concerned with the biological mediation and control of behavior. I have been selective in deciding what physiological mechanisms to discuss and throughout have attempted to integrate a discussion of physiological mechanisms with ethological and experiential approaches. In some instances, such as in understanding the regulation of feeding behavior and appetite or in analyzing different categories of aggressive behavior, it is useful to point out what parts of the central nervous system are involved in controlling behavior. In other instances, I have stressed hormonal control because for some behaviors, hormones are the most easily manipulated physiological variable. Certain behavioral experiences such as stress or confinement may affect hormone secretions, which in turn alter an organism's physiology. When hormones are altered directly through castration, certain behavioral patterns are altered rather remarkably. Because of our direct control of the environment of domestic animals and the physiological alterations we subject them to, some emphasis on the physiological approach to understanding behavior is necessary.

References

CLUTTON-BROCK, J. 1977. Manmade dogs. *Science* 197:1340–1342.

DAVIS, S. J. M. AND F. R. VALLA. 1978. Evidence for domestication of the dog 12,000 years ago in the Natufian of Israel. *Nature* 276:608–610.

DOWNS, J. F. 1960. Domestication: An examination of the changing social relationship between man and animals. *Kroeber Anthropological Society Papers*, no. 22, Spring, 18–67.

HALE, E. B. 1969. Domestication and the evolution of behavior, In *The behaviour of domestic animals*. 2d ed., pp. 22–44, ed. E. S. E. Hafez. Baltimore: Williams and Wilkins.

HART, B. L. 1978. *Feline behavior.* Santa Barbara, Calif.: Veterinary Practice Publishing Co.

HART, B. L. 1980. *Canine behavior.* Santa Barbara, Calif.: Veterinary Practice Publishing Co.

ISAAC, E. 1962. On the domestication of cattle. *Science* 137:195–204.

ISAAC, E. 1970. *Geography of domestication.* Englewood Cliffs: Prentice-Hall.

JARMAN, M. R. 1976. Early animal husbandry. *Phil. Trans. R. Soc. London*, Ser. P., 275:85–97.

KRATZER, D. D. 1971. Learning in farm animals. *J. Anim. Sci.* 32:1268–1273.

LAWRENCE, E. A. 1982. *Rodeo: An anthropologist looks at the wild and the tame.* Knoxville, Tennessee: University of Tennessee Press.

MC KNIGHT, T. 1964. Feral livestock in Anglo-America. In *University of California Publications in Geography*, vol. 16, pp. 1–87. Berkeley: University of California Press.

PROTSCH, R., AND R. BERGER. 1973. Earliest radiocarbon dates for domesticated animals. *Science* 179:235–239.

TENNESSEN, T., AND R. J. HUDSON. 1981. Traits relevant to the domestication of herbivores. *Appl. Anim. Ethol.* 7:87–102.

UCKO, P. J., AND G. W. DIMBLEBY. 1969. *The domestication and exploitation of plants and animals.* London: Duckworth.

ZEUNER, F. E. 1963. *A history of domesticated animals.* New York: Harper & Row.

CHAPTER TWO ▪ SOCIAL ORGANIZATION, COMMUNICATION, AND AGGRESSIVE BEHAVIOR

Domestic animals are usually maintained in groups. The groupings may range from two or three dogs, cats, or horses kept as pets to huge herds of livestock being finished for market or bred for range production. Whenever animals are grouped, they interact with one another, often aggressively but also in friendly or amicable ways. Social interactions are coordinated by communication. The range of behavioral patterns involved in the interactions and communication of domestic animals is the subject of this chapter. The term social behavior often includes a behavioral chain of reciprocal interactions between or among animals of the same species. The wild ancestors and relatives of the domestic species display a number of innately controlled, species-specific social behavioral patterns that have proved to be highly adaptive in the natural setting.

Some of the more interesting behavioral patterns occur during conflicts. Conflicts may be settled by threats or by fighting. The term agonistic behavior refers to the species-typical threats and submissive gestures, as well as actual fighting, which resolve

conflicts. The term aggressive behavior is usually used interchangeably with agonistic behavior, and we shall do so here. In
a strict sense the term refers to attack and actual fighting.

Although social behavior usually concerns the relationships
between animals of the same species, on the domestic scene
social interactions occur between animals and their human owners or between different animal species such as sheep and goats.
Some of the interspecific emotional attachments may be as
strong as those within a species. Such attachments have occurred, for example, between dogs raised in human families and
their owners.

Domestic animals are maintained in groups that are usually
much different than those of the wild ancestors. This is important to keep in mind when we consider social structure, especially dominance hierarchies. The social structure we see in artificial grouping can give us erroneous notions about the social
structure that might exist in feral, or wild, populations.

What are some of these differences in the makeup of groups
in the domestic and wild populations? Sometimes we maintain
animals in groups of only two or three so that the possibility of
social interactions between young and old, and between males
and females, does not exist. On large farms we tend to group
the animals that are uniform in size and sex. Males are castrated,
which reduces the aggressive activity. The groups may be excessively large to the point where individual recognition among
animals is impossible. Grouping in domestic environments also
differs because we, as handlers, interact with animals by forcing
ourselves into their social hierarchy, using threats or devices to
win aggressive encounters.

Even with these major modifications of social interactions,
domestic animals display many aspects of the social behavior
characteristic of their wild ancestors. Since we must raise animals in groups, and interact with them, it is important to understand the various aspects of social behavior. The better social
behavior is understood and dealt with, the more successful we
can be in animal management.

Species that are predisposed to grouping as a reflection of
grouping in their wild ancestors are referred to as social species.
The nature of this sociality is evident when an animal is taken
from its group. Animals may vocalize excessively. Goats, for
example, rear up a great deal when isolated, even more than do
sheep. Price and Thos (1980) attribute the rearing tendency of

goats to a possible adaptation of locating members of their group among trees and shrubs, which is more their natural habitat than grassland (Huston, 1978).

In domestic animals we have produced distinct breeds through selective breeding and there is evidence, at least in sheep, that individuals of a breed prefer to socialize with members of their own breed (Arnold and Pahl, 1974). Animals reared with members of a different species—sheep with goats, lambs with dogs, monkeys with dogs—develop social affinities for the species with which they were raised (Cairns and Johnson, 1965; Tomlinson and Price, 1980).

Within the group there is a social organization that usually reflects dominant-subordinate relationships. Social interactions within relationships range from dramatic aggressive encounters to the more typical amicable interactions, and sections of this chapter treat both of these extremes. Play, a special type of amicable behavior, usually constitutes a high proportion of the social interactions between infants and juveniles. The rich behavioral repertoires associated with sexual and maternal behavior are each dealt with in separate chapters.

Species in which the individuals do not naturally form groups as adults are considered solitary or asocial. The only domestic species with an asocial ancestor is the cat. Of course, asocial animals also interact, at times, with conspecifics, so that communication, territoriality, and agonistic interactions are a concern. All species, whether social or not, must come together for reproductive purposes. Thus, males may compete for females and eventually males interact with females sexually. The young offspring of a mother of an asocial species live in a social group with littermates before the offspring separate for more solitary existences.

■

AGGRESSIVE BEHAVIOR

Social interactions between animals often involve some degree of conflict. This may be related to obtaining food, sexual mates, or preferred rest areas. Other sources of conflict arise as a result of a tendency of animals to protect a territory or offspring. Sometimes the cause of the conflict is simply the presence of another animal.

The behavioral patterns related to conflict are referred to as agonistic behavior. This includes threat, flight, and submissive behavior as well as actual attack. The term aggressive behavior, in the narrow sense, is reserved for behavior of a type that could cause physical injury to another animal. With regard to domestic animals, it may be best to use a little more liberal interpretation of the term aggressive behavior to include the more intense forms of threat, when it is evident that an attack is imminent. Our reference to an aggressive individual, or an aggressive breed, usually points to the ease with which some animals are provoked to attack. Most of the conceptual thinking about agonistic behavior involves interactions between members of the same species. Agonistic behavioral patterns are especially important in establishing and maintaining dominant-subordinate relationships among social species. A special type of interspecific aggressive behavior is seen when a predator stalks and attacks its prey. Aside from predatory aggression, agonistic behavior may be interspecific when a member of one species enters the social order of another. Humans seem to commonly enter the social order of dogs, horses, and cattle. Threat, attack, submission, and flight reactions may be exhibited by an individual of one species toward a member of the other species. The nomadic Fulani herdsmen of northern Nigeria, for example, enter into the social system of their cattle and assume the role of the dominant in order to control their movement without the use of ropes, fences, or corrals (Lott and Hart, 1979). Although the herdsmen cannot display the same forms of physical aggression as herd bulls, they maintain their position through consistent threat behavior (Figure 2-1). Occasionally a goat may enter the social order of a flock of sheep, or a household cat may appear to enter the social order of one or two dogs in the same house.

Behavioral Patterns of Threat and Submission

The type of social structure characteristic of a species determines the types of agonistic interactions that have evolved. Solitary or asocial species such as cats, which do not establish clear dominance hierarchies as readily as social species, tend to fight much more frequently than social species. A good illustration of this is that free-roaming, gonadally intact male cats have a much higher frequency of fight wounds or abscesses than

Figure 2-1 A nomadic Fulani herdsman of northern Nigeria controls his cattle and keeps them from grazing in a field of cultivated corn by threats. (From Lott and Hart, 1979.)

free-roaming male dogs because cats settle conflicts more often by fighting rather than by threat and submission.

Whether they are members of a social or solitary species, it is still unusual for two animals who happen to come upon each other to immediately engage in a vicious fight. Threats almost always precede fighting. For an animal that is obviously smaller or weaker and is away from its territory, threat by a rival is often sufficient to drive it away or evoke submission. Similarly, if there has recently been a fight between two animals with the defeat of one, the previously defeated animal may readily withdraw from a new encounter or indicate submission.

If a fight does ensue, it is generally brief, with the weaker or less aggressive individual usually being allowed to escape. In the social species the flight of the defeated animal or the display of a submissive posture often has a direct inhibitory influence on further aggression by the stronger of the two.

Threat Behavior   The behavioral aspects of threat, submission, and fighting have been elaborated upon by ethologists. The approaches are oriented toward wild species, but many of the concepts apply to domestic animals. There are two basic types of threat behavior. One type includes those behavioral patterns indicating that the animal is likely to attack if its rival does not move or leave; this is the offensive threat. The second type, the defensive threat, involves those patterns indicating that the animal will not attack unless the rival persists in coming too close (Ewer, 1968). Often different postures or facial grimaces are used for offensive and defensive threats.

The postures and movements assumed during the threat generally indicate the intention of the animal in regard to the next movement that is most likely (Figure 2-2). In an aggressive threat, a dog may, in addition to snarling, growling, and holding its ears down, push its head and neck forward and position the limbs in such a way that the animal is ready to move forward. In the defensive threat, on the other hand, the dog may turn sideways to its rival or even lie on its back while at the same time snarling and growling.

Cats that are ready to attack use a posture from which they can spring into action. In the defensive threat they turn sideways to the opponent with the back arched. This is the classic Halloween cat picture; in this posture the animal is in no position to attack.

Some threats tend to reveal, or even exaggerate, characteristics such as the size, strength, or weapons of an animal. In dogs, snarling results in baring of the large canine teeth. At the same time piloerection occurs over the shoulder region, enhancing the apparent size. Piloerection over the neck and shoulders is also seen in threat postures of goats. Lowering the ears is seen in most domestic animals as part of aggressive and defensive threats.

A threat posture often observed in wild bovids and domestic cattle is the broadside threat, in which the animal turns its body almost at right angles to its opponent but with the head, neck, and shoulder toward the opponent (Figure 2-3). Bulls may either withdraw or attack from this posture (Fraser, 1957). The broadside threat is also displayed to unfamiliar people by bulls. Another threat behavior of bulls is pawing of dirt, rubbing the head and neck on the ground, and digging the horns into the ground (Figure 2-3).

Threat behavior of pigs consists of emitting barking grunts and champing the jaws together along with grinding the teeth while producing an abundant flow of saliva. The saliva turns to froth and falls in clumps on the ground.

Figure 2-2 Increasing intensity of threat by a female goat: The eye stare displayed by the dominant doe (top, right) to a subordinate may displace the subordinate. Intention movements with horns (middle, right) provide the next level of threat if the eye stare does not displace the subordinate. Mild contact with horns is the next level of threat that falls short of an actual butt, and usually succeeds in displacing the subordinate (bottom, left).

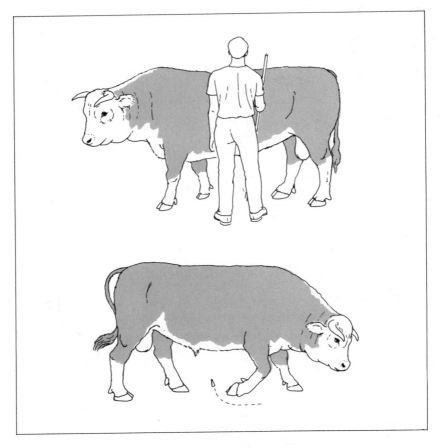

Figure 2-3   Threat postures of bulls: The broadside threat (top) may be directed toward people or other bulls. The dirt pawing (bottom) is often displayed in a pasture at certain sparring locations; the pawing of dirt or digging in it with the horns serves to invite other bulls in the pasture over for a fight.

Growling is a part of threat behavior in a large variety of mammalian species. In dogs and cats it may constitute part of the defensive or offensive threat.

Some aspects of offensive threats in domestic animals are most prominently displayed by gonadally intact males. Female cattle, for example, do not have the same tendency as males to paw the earth. In swine threatening by females does not include jaw champing and saliva production. On the other hand, female dogs and cats both display the prominent piloerection when threatening.

Figure 2-4 Example of the role of inhibition in fighting: Rearing by one billy goat tends to provoke the other to rear (top), but if one opponent does not rear, the other will usually not deliver a blow (bottom).

Submissive Behavior One of the factors in the evolution of submissive behavior is that it is not often feasible for an animal to withdraw from an opponent by leaving the group. One of the frequent observations of animal social behavior is that a submissive response acts as an inhibitory stimulus to the continuance of threat or aggression by the opponent. Thus, the submissive animal is spared serious injury, and the stronger animal gets its way without further risk of injury (Figure 2-4).

Many submissive patterns are visual displays. Diversion of eye contact is a frequent indication of submission among canids and equids. When the dominant animal looks directly at the subordinate, the subordinate diverts its eyes and sometimes the head. The role of eye diversion is often evident between male dogs and their owners. If the owners are dominant, looking directly into the dog's eyes evokes eye and head diversion.

The submissive posture of creeping along with the tail tucked between the back legs is displayed by dogs when they wish to approach a superior one even though the superior may give no threat or show any intention of attacking. This is frequently seen in puppies who approach adults. Females may occasionally use the same behavior to approach males or human owners. The belly-up submissive posture displayed by puppies to adult dogs and people, and often by adult females to people, is commonly recognized (Figure 2-5). With these more extreme forms of submission, the submissive animal is completely at the mercy of the other animal but is very rarely attacked or harmed.

Puppies often release a small amount of urine when displaying submissive behavior. The dominant individual, if it is an older dog, sometimes responds by licking up the puppy's urine, thus seemingly converting the interaction into a parent-child relationship (Ewer, 1968). Adult male dogs rarely show the belly-up posture as a sign of submissiveness. Submissive behavior for adult male dogs toward other dogs or toward people is a crouching stance with the head lowered and eyes diverted. There may even be some growling, which suggests that the inferior animal, while admitting its subordinate position, is grumbling about it.

A form of submissive facial expression in horses has been observed in response to threats by older animals in which the submissive animal makes nibbling movements without actually closing the jaws. The corners of the mouth are not drawn back as in threatening and the ears are turned so as to project laterally and downward.

Another type of submissive behavior is vocalizations, such as the yelp of dogs, the squeal of pigs, and possibly crying by humans. The most extreme form of submission, according to some ethologists, is the death feign, which is so characteristic of the Virginia opossum and also of a species of fox (Ewer, 1968).

Fighting Behavior

Although wild animals may fight conspecifics to the death, fighting has impressed most behaviorists by its generally harmless effects on the participants. As was pointed out in the previous section, on withdrawal of one animal or the display of a submissive posture, the opponent does not normally proceed to

Figure 2-5 Forms of submissive behavior in dogs: The posture shown in the middle, with the ears back and head down (left), is more pronounced than simple eye aversion and avoiding the dominant animal (top, left), but is less extreme than the supine posture (bottom, left), which is more characteristic of juveniles.

kill or seriously injure the other. One might wonder about the adaptability of this inhibition in wild animals since the defeated animal, particularly if it is younger than the winner, is likely to offer competition to the winner at some future date. However, the inhibition does conserve the energy of the stronger animal and removes the danger of its being wounded while winning. Ending up with an infected wound may be as bad as losing.

To observers most fighting seems to be characterized by be-havior that avoids serious damage to the opponents but still tests the strength and determination of the two animals, indicating who the winner would be were the fight to be pursued to the end. Thus in a fight between dogs it is evident that even in what looks like a ferocious battle, the biting that the animals often mete out is inhibited and not nearly as severe as that used in attacks on prey. From the dog's standpoint these may be quite inhibited. Of course, people are frequently the target of dog biting. However, to us, even an inhibited bite is painful, some-times dangerous, and usually worthy of complaint.

Among the ungulates, especially those that possess special weapons such as horns, antlers, or tusks, we see a conservative type of fighting that greatly reduces the chances of injury. Goats fight, for example, by rearing and then forcefully butting each other just as they hit the ground (Figure 2-4). Sheep fight by backing off from each other and then rushing toward each other to ram horns (Figure 2-6). In both instances the head is used to absorb the shock. If the same intensity of butting hit other parts of the body, the animals could easily be seriously injured.

There is a common erroneous feeling that horns and antlers evolved as weapons for defense against predators. However, al-though hornlike organs may occasionally be used against pre-dators, there is clear evidence that they did not evolve under this influence. Antlers are functional for only a short time in the year, and in fact deer lose their antlers in the winter, when they tend to be most harassed by predators. Deer are well able to defend themselves with their forelegs and hindlegs. Many ungulates such as sheep or goats can best protect themselves by flight rather than by using their horns (Geist, 1965).

Hence the main value of horns and antlers is for aggressive contacts between members of the same species. Mammals of fairly large size probably developed a tendency to fight with the head because the fast, vigorous movements characteristic of smaller species were difficult or dangerous. Thus, the appear-ance of horns is considered to be associated with the behavior of delving out blows with the head and a reduction of fighting that involves more of the body. The same can be argued for the development of large tusks in pigs.

The best defense for some species with horns is to catch the opponent's blow with a counterblow and thus make it ineffec-tive. This is well exemplified in the fighting behavior of goats

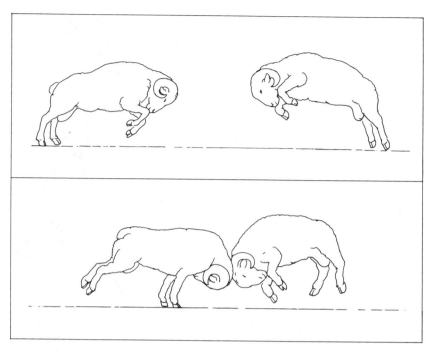

Figure 2-6 Stylized fighting in rams: The animals may go at each other with tremendous force, as shown just before and after contact in these drawings from motion picture frames. There is usually no injury.

and sheep. The skull is thick, with large sinuses forming a double roof over the brain, which allows the head to withstand very great ramming forces. In animals in which butting or ramming is not a form of fighting, the skull is much thinner and the sinuses smaller. Cattle horns are also curved outwards so that the animals can hook the horns of an opponent when they are directed toward the chest or abdomen. If an animal misses catching the opponent's horns, puncture wounds or slashes may result. The spirals, bumps, and ridges on the horns of various antelopes and gazelles, which look so dangerous to human observers, serve to lock the horns together, thereby preventing slippage to the side. This then turns a batting contest into a full-strength wrestling, pushing, and shoving match.

There is a clear sexual dimorphism in fighting styles in goats and sheep. Males display the behavior depicted in Figure 2-6. Females exhibit more variability in their fighting style and may deliver blows to the body of the opponent as well as to the head.

Sexual dimorphism in fighting is not so obvious on other species. In cattle one animal strives for a flank attack on the other. The opponent attempts to force the attacker to a frontal attack and, if necessary, backtracks to avoid being butted in the flank. A frontal butting and pushing encounter may last to the point of physical exhaustion of the animals.

In animals that have horn-like weapons the typical fighting appears to be stylized or ritualized, and in fact the contests seem to follow a set of rules. These observations, along with others on various species of birds and fish, have given rise to the school of thought that animals fight by a set of evolved behavioral rules such that opponents avoid injuring each other, while still determining which animal is strongest (Lorenz, 1966). Some specialists in animal social behavior have difficulty with this notion since opponents need only break the rules to reap an immediate advantage in winning the fights. Rather than assuming that animals are abiding by an inherited set of rules for fighting, an alternative explanation is that an animal's defensive posture, needed to prevent injury to itself, puts constraints upon its offensive movements, much as a human boxer must deliver blows from a protective stance. The result could then be the evolution of a form of fighting which looks ritualized but is as vicious or fierce as possible under the constraints of an adequate defense posture.

In pigs fighting is fairly similar among males and females, except that sows attempt to bite an opponent and boars attempt to slash an opponent with their tusks. The canine teeth of domestic pigs are usually cut in infancy to prevent injuries during fighting. Nonetheless, the pattern of fighting is such that serious damage to the rivals does not occur even if the animals have fully developed tusks. Pigs begin fighting by pushing shoulder to shoulder and eventually end up striking each other with the head and tusks. Although the shoulders of one or both animals may become lacerated, the thick skin over the shoulder is generally sufficient to minimize the damage. Boars that are fighting seem to look for an opportunity to bite a front leg or an ear of the opponent. Occasionally they may back off and charge the side of the opponent with the mouth open, ready to bite (Hafez and Signoret, 1969). Most fights end rather quickly, but the dominant boar often pursues the loser and bites or slashes him.

Although tomcats do often injure each other, the pattern of fighting is such as to limit serious injury. Actual biting is uncommon, and most of the battle is fought with claws. The postures frequently assumed, with the animals rolling on their backs and using the claws of all four feet, allows neither animal much opportunity for the use of the teeth. Along with claws, vocalizations appear to play an important role. As many cat owners have noted, the amount of noise is out of proportion to actual damage that is being done (Ewer, 1968).

Physiological and Environmental Aspects of Aggression

Internal Causes of Agression One of the issues that is obviously of concern in regard to human aggressive behavior is the degree to which aggression may be viewed as a biological drive in which deprivation increases the likelihood of fighting. The concept of the inevitability of aggression has been popularized by Lorenz (1966), Ardrey (1966), and other writers. It is suggested that a type of biological energy builds up in an individual to the point where it must eventually be released. This model was originally proposed from observations of birds and fishes in which the tendency for some animals to attack an inappropriate stimulus, such as a dummy, develops if more natural aggressive attacks are prevented.

This hydraulic energy model, which assumes that an aggressive drive accumulates with time and activates attack behavior, applies to only a very few species and has virtually no support from the physiological standpoint. However, the model still retains its popularity even as applied to the human species because of the widespread occurrence of aggression in animals and throughout human history.

Although aggressive behavior does not seem to be a manifestation of an inevitable biological drive that accumulates over time, fighting, as well as other aspects of agonistic behavior, consists of stereotyped, species-typical patterns with a definite neurological basis. Studies have shown, for example, that in the cat stimulation of the lateral hypothalamus of the brain will evoke an attack on a rat with the subject ignoring the experimenter, whereas stimulation of the medial hypothalamus evokes an attack on the experimenter, and the cat ignores the rat (Flynn, 1967).

In terms of the biological basis of aggressive behavior it appears reasonable at this time to assume that there is a genetically programmed neural substrate of aggressive behavior, which may be readily evoked or triggered by environmental stimuli but is not under a neurological energizer that inevitably activates it.

Environmental Factors Influencing Aggressive Behavior   There are numerous factors or stimuli in an animal's environment that tend to evoke aggressive behavior. Those that we shall give particular attention to are the proximity to other animals, painful stimulation, and events that may be considered frustrating.

The closer animals are crowded together, the more likely are aggressive attacks. This was illustrated in a study of pigs showing that decreasing the area available per pig in an enclosure increased the number of aggressive interactions as well as the severity of attacks (Ewbank and Bryant, 1972). In many species, isolation prior to placing animals together tends to increase the probability of fighting.

Sex also plays a role in aggressive behavior in that males of almost all species have a much higher tendency to fight than females. This is a common observation among domestic animals, and it is supported by field and laboratory studies of other animals (Hart, 1974b). Aggressive behavior between males is so prominent that it is referred to as intermale aggression. However, females engage in agonistic interactions, and sometimes females may be as brutal as males in their aggressive attacks on subordinates.

Painful stimulation, or even the fear of pain, has long been recognized as inducing aggressive behavior. This is evident in watching puppies playing and being able to predict the almost inevitable fight that ensues as some individuals start to play too roughly and hurt the others.

In natural situations during fighting one animal inflicts pain on another, and this in turn tends to induce the receiver to fight more intensely. Eventually, of course, one animal may inflict a sufficient degree of pain on its opponent that the opponent withdraws.

Historically, one of the more popular lines of research in aggressive behavior involved the frustration-aggression hypothesis. This concept seems particularly appropriate for human aggressive behavior, where scapegoating is observed. Interestingly, one of the best examples of scapegoating in animals was observed in a herd of goats. When food was restricted, hungry goats, frustrated by the attempts of other animals to prevent them from obtaining food, became very aggressive toward subordinates, whereas they normally acted peacefully (Scott, 1948).

The earlier formulations of the frustration-aggression hypothesis were quite circular in claiming that when aggression occurred frustration was obviously the cause, and when a situation seemed possibly frustrating, but no aggression occurred, that the situation was not actually frustrating. More recent considerations of frustration have emphasized that frustration probably only increases the likelihood of aggression and, on the other hand, aggression is only one of the outcomes of frustration (Hinde, 1969).

It is clear that the proximity of animals, pain, or frustration do not invariably, or even usually, lead to aggression. All of the factors interact. Thus frustration and the occurrence of painful stimuli may produce aggression when neither factor acting alone would.

Roles of Hormones and Age in Intermale Fighting   As mentioned previously, males generally have a stronger tendency to fight each other when placed together than females. There is even less of a tendency for males to fight females. This difference in aggressiveness between males and females represents a sexual dimorphism that is undoubtedly related to the secretion of testosterone in males during the perinatal period. In the adult, aggressive behavior is also related to testosterone secretion, and castration of adult male animals reduces fighting activity (Hart, 1974*b*). Interestingly, there are species differences in the effectiveness of this operation. Castration, for example, reduces fighting more frequently in adult male cats than adult male dogs (Hart and Barrett, 1973; Hopkins et al., 1976).

Related to the secretion of testosterone is the factor of age. Young animals of all domestic mammals engage in play fighting. As animals mature play fighting becomes less frequent and fighting, when it does occur, is serious. This change, which is seen in both males and females, is a function of maturation.

Types of Aggressive Behavior

It is perhaps well to summarize the discussion of agonistic behavior by considering the different types of aggressive encounters that may occur in animals based on the antecedent causes or stimuli. Moyer (1968) has emphasized that aggression is not a unitary phenomenon and has suggested the following classes of aggressive behavior: predatory, intermale, fear-induced, irritable, territorial, maternal, and learned. Some modification of this classification, especially for domestic animals, seems appropriate, and the categories of aggressive behavior I shall use are predator–prey, intermale, competitive, fear-induced, pain-induced, territorial-social, maternal, learned, and idiopathic.

Predator–Prey Aggression   Inasmuch as agonistic behavior mostly relates to the interactions between animals of the same species, predator–prey aggression should be considered apart from the other types of aggressive behavior. The behavior both of the predator and the prey may differ considerably from the other types agonistic behavior. For example, wolves or dogs do not use the inhibited bite in attacking prey as they do in fighting each other. Deer use their hooves against predators such as wolves, whereas antlers are used in fighting each other.

A cat uses a bite with the canine teeth in attacking prey but usually the claws in fighting other cats. In stalking small rodents or birds, the cat approaches with the body close to the ground and may periodically pause or attempt to hide under some type of cover. When the cat is close enough, it pounces and grabs the prey with outstretched claws and teeth. The prey, which is killed or wounded by being bitten in the neck, is often carried to a secluded place and eaten (Leyhausen, 1979).

Most cat owners have seen cats playing with small rodents, repeatedly going through the responses of stalking, pouncing, grabbing, and shaking the hapless rodent without killing it. This appears to be a form of playing and could be a manifestation of the fact that domestic cats have a limited opportunity to play (Ewer, 1968). Once the prey finally dies, or collapses from exhaustion, eating may be initiated by the lack of motion of the prey. In cats, predatory behavior will occur even though the animal may not be hungry. In fact, if one is interested in keeping cats around for rodent control, feeding them regularly will keep them attached to the area.

Intermale Aggression   Males of many species, including all of the domestic species, have a special propensity for spontaneous fighting that is uncommon in females. This type of behavior seems to reflect a sexual dimorphism in the neonatal development of the brain, and even administration of testosterone to females (adults) is relatively ineffective in making females more aggressive. The behavior is androgen-dependent in that intermale fighting does not occur until the pubertal surge in testosterone secretion has occurred. Prepubertal castration usually prevents this behavior from occurring. This is one of the reasons cats, horses, and all livestock species are castrated prepubertally. In cats, dogs, and horses, postpubertal castration reduces the likelihood of intermale fighting.

The study of this type of aggressive behavior has given rise to the classical descriptions of threat displays and submissive postures. It is evidently involved in the establishment of dominance hierarchies among males in social groups. However, dominance hierarchies are also rigid in female groups, so that intermale aggression cannot be taken as the sole basis for establishing and maintaining dominance hierarchies.

Competitive Aggression   An animal may contest and fight for food, a preferred resting spot, or for dominance over another animal. A specific name for the latter behavior is dominance-related aggression. Like intermale aggression, this type of behavior involves threats and submissive postures, and in a social group the agonistic reactions usually do not reach the point of aggressive attacks. It may be difficult to distinguish this type of aggression from intermale aggression when the individuals involved are both males; actually the aggression may be a function of both factors. In competitive situations, males may also behave agonistically toward females and certainly females may be as aggressive toward each other as males. Of course human handlers must establish dominance over their animals. On occasion they may be the target of aggressive responses from the animal they are attempting to dominate.

Fear-Induced Aggression   This type of aggression may be displayed by an animal that would escape if it were possible but is prevented from doing so. Often it may threaten and then attack its assailant; the threat is usually of the defensive type. This aggressive behavior, of course, may be exhibited toward a member of the animal's own species or different species. It is probably the most common type of aggression directed toward humans by animals. The behavior is often self-reinforcing in that the growling, threats, or biting drive away the people responsible for inducing the fear reaction. There is no evidence that castration alters this type of aggressive activity. Fear-induced aggressive behavior occurs in wild animals maintained in captivity. In zoo animals tranquilizers have been effective in reducing this behavior.

Pain-Induced Aggression   Both males and females have an innate response to attack if painfully stimulated. Pain-induced aggression does not ordinarily present a serious behavioral problem. However, we should be aware of the fact that in attempting to break up a fight by painful stimulation such as hitting one or both dogs, or even using an electronic shock collar, we may intensify the fight.

Territorial-Social Aggression   Most domestic species will act aggressively toward intruders placed in their pens or home areas. Male and female pigs are especially vicious in their attack of new animals introduced into their group. Goats, horses, and cattle also attack new individuals, although injuries are less frequent. Dogs and cats are often aggressive toward strange animals entering their territory. Dogs often identify people, as well as other dogs, as a threat to their territory; thus they make ideal guards for a house and yard. In addition to threatening or attacking intruders, dogs use a specific alarm bark to arouse the family as it would a pack. Dogs are usually indifferent to territorial intrusions by other animals, such as cats, birds, horses, or cattle.

Animals such as livestock species, although not strongly attached to a particular territory, still act aggressively toward strangers to their group; thus they act to protect a group in somewhat the same manner as territorial animals protect a home area. Animals that invade a territory or social group of others have long been recognized as having a low probability of winning an aggressive encounter.

Maternal Aggression   Females of most species of wild animals are capable of intense aggression in defense of their young. This is one of the few types of aggressive behavior in which an animal may attack another with no threat display. This type of aggressive activity is believed to be a function of the hormonal state of the female during lactation as well as the presence of the young.

In the domestic species, breeding and husbandry practices have resulted in a reduction of this behavior because we have selected against it in order to play a role in raising the young. Females retained for breeding have been those with the least tendency to show this behavior. Of the domestic species, swine seem to exhibit the most intense form of maternal aggression.

Learned Aggression   Although learning influences all of the categories of aggressive behavior, there is a special type in which some animals, particularly dogs, may be trained to attack. This is the type of aggression displayed by guard dogs that attack on command.

Idiopathic Aggression   This is a catchall category for abnormal aggression such as that which may be related to brain injury. Also included is self-directed aggression stemming from adverse early experiences. A third type of idiopathic aggression has been described for dogs. A typical case is the dog that the owner describes as almost always friendly, easily controlled and well mannered, but that, for no explainable reason, may suddenly turn and viciously attack a member of the household or a friend that the dog knows. The behavior is unprovoked and even unpredictable. Often the dog gives very little warning before attacking, and the attacks may be directed toward the person's face or neck. Owners of such dogs have frequently mentioned that when their dogs do attack they no longer seem to recognize the people in the family and may even get a glazed or distant look in their eyes. The breeds in which this behavior has become a noticeable problem, namely Saint Bernards, Doberman pinschers, Bernese Mountain Dogs, and German Shepherds, are so large as to pose a real danger (Hart, 1977b).

Aggression Toward People

For anyone handling animals it is obvious that familiarity with threat displays and fighting behavior is important. The aggressive tendency of many animals may be related to a critical distance that, if entered, may evoke agonistic or attack behavior. The first sign of threat gives a notice that the critical distance is being approached. Animals often learn the responses of human handlers, and if they perceive signs of nervousness their threat may turn into an attack. A person who approaches the critical distance with an air of dominance may be met by a threat but no attack may occur, and some degree of submissiveness may be displayed.

Most of the social animals treat humans as a conspecific rather than as a predator in terms of agonistic behavior. Humans are usually in the dominant role, and most animals seem therefore to avoid initiation of fights. Most attacks on humans by animals are fear-induced. In some instances, especially when confronting strange animals, a threat by an animal could be followed by an attack. A submissive response by the person, such as diverting the eyes, turning away, or backing off may prevent an attack.

I have noted in the earlier discussion that painful stimulation often tends to evoke or stimulate fighting. This is very evident in handling cats. Restraining the animal or administering punishment is likely to evoke aggression. It is almost impossible to subdue most cats by force or pain.

Although it is rather pointless to attempt to control cats by aversive stimulation, such stimulation can be of real value in dealing with dogs or horses. Punishment, apart from sometimes altering undesirable behavior, helps to establish and maintain the dominant position of the human handler. Painful stimulation may produce submission and facilitate restraint of an animal. This is clearly evident in the use of the lip chain, or twitch, in controlling horses. Such animals obviously could still physically overpower the handler, but most horses become subdued with this stimulation. If in response to punishment an animal becomes quiet and submissive, then as the relationship between the dominant person and the animal develops, the use of actual painful stimulation often becomes unnecessary and just a certain tone of voice and gesture are enough to evoke a submissive response.

■

ANIMAL COMMUNICATION AND IDENTIFICATION

Social interactions between animals require communication. This involves a sender and a receiver of a signal. Although the sender may not use specialized organs to initiate a message, the receiver usually has sensory organs specialized for the reception of the different sensory modalities. The channels or modalities used in communication among domestic animals are sound, visual displays, contact, and chemical substances. Redundancy in the use of communication modalities is very common and extends the gradations that can be expressed. For example, a dog that is threatening another dog may display a snarl, but in addition to this visual modality, it often uses the auditory modality of growling as well.

One advantage of the simultaneous use of two signals, such as a visual and an auditory signal, is the communication of two kinds of information (Marler, 1967). For example, if the acoustic signal is an alarm sound, this alerts the receiver and gives information about the presence of a predator. By noting the direction to which the signaler is looking, the receiver also receives information about the location of the predator.

A number of behavioral functions can be ascribed to communication. These functions involve locating and identifying others, issuing commands, reinforcing social status, and indicating a momentary mood.

Social interactions also require identification of sexual and social status, physiological state, colony versus noncolony membership, and estrous versus diestrous breeding condition, and these interactions may include individual recognition. In identifying individuals visual, auditory, contact, and olfactory stimuli are important. Animals may recognize one another by color or hair coat patterns and body shape. Young are differentiated from adults on the basis of size. Animals can recognize differences in vocalizations (Poindron and Carrick, 1976). Perhaps the best example is the ability of ewes and lambs to recognize each other by particular vocalizations. Recognition by physical contact is limited to the more intimate relationships such as that between mother and infant. Olfactory stimuli are extremely important in individual and subgroup recognition in mammals, and this topic is dealt with in some detail below.

Since identification and communication involve the same sensory modalities, and in fact the two processes overlap, we shall deal with both processes together under each modality.

Vocalizations

Most animal sounds are quite obvious to us. Usually vocalizations of domestic animals are within the human auditory range, although most animals can hear sounds above the upper frequency threshold of the human ear (Brown and Pye, 1975). In the nest newborn rodents emit ultrasonic calls as a means of communicating with the mother (Geyer, 1979). Ultrasonic vocalizations also attract adult male and female rats to each other during sexual encounters as well as regulating distance between the sexes following copulation (Geyer and Barfield, 1980; Geyer et al., 1978).

The extent to which ultrasound is used in domestic animals is not known. Not all animal sounds, of course, involve communication. Sounds produced in the oral cavity resulting from panting or eating do not involve communication.

The production of sound gives an animal the advantage of communicating but leaving the body free for other activities. In the wild it minimizes the degree to which the signal can be used by predators for tracking because of the rapid fading characteristic of sound. Sound also has the property of flexibility in amplitude, frequency, and time.

Animals display a variety of vocalizations. The functions of some of them are quite clear (Kiley, 1972). Mother sheep, cattle, pigs, and cats have readily recognizable calls that attract their young. Pigs, dogs, and horses have alarm sounds. The alarm bark tendency in dogs is the main reason some dogs are kept as pets. The sex call of the female cat is used to attract males in her territory. The boar's mating vocalization contributes to inducing the sow's mating posture (Signoret et al., 1960). And in refractory bulls, mating can be accelerated by the sound of a cow's bellowing (De Vuyst et al., 1964).

The variety of species-specific animal sounds is too numerous to mention here (Sales and Pye, 1974; Sebeok, 1977). There are vocalizations directed by adult animals to each other and vocalizations exchanged between young and their mothers. The latter are the least complicated. Kittens, for example, emit an intense vocalization within seconds after being removed from their mothers. Mother cats emit a low-intensity gargly sound when their kittens are taken away. This sound may be a peremptory signal for the kitten to return or a guiding stimulus for the kitten to follow. Adult cats give out a meow customarily heard just prior to our feeding them. They may also emit an intense call when hurt or stressed. Cats may also growl and hiss, crouching, and staring. The last is the more extreme threat, made usually just prior to striking (Brown et al., 1978).

A lengthy catalogue of vocalizations has been worked out for canids. In the coyote Lehner (1978) describes a growl, huff, woof, bark, bark-howl, whine, wow-oo-wow, yelp, lone yowl, group howl, and group yip-howl. Each of these sounds has a particular communicatory function. Most of these vocalizations are also heard in the wolf and dog (Bleicher, 1963; Mech, 1970). Lehner mentions there is evidence that coyote bitches may use an ultrasonic vocalization to call their pups from the den.

An interesting type of vocal communication familiar to dog owners is howling. The function of this behavior has become clearer in studies on the wolf. Howling appears to play a role in intrapack identification, in expediting and coordinating departures, reunions, and movements. It may also mediate some interpack communication, primarily avoidance of one pack by another (Harrington and Mech, 1978).

Visual Displays

This mode of communication is about as common as sound in domestic animals (Figures 2-7 and 2-8). Threat postures such sas that of a horse with ears laid back communicates social status or an aggressive command for another animal to move. With the exception of special warning signals, most visual displays are closely related to a predisposition toward a certain response. For example, the receptive posture of an estrous female cat that communicates her receptivity is involved in copulation. Pushing the head forward and lowering the ears are aspects of the attack posture; they are used in offensive threats. Examples of visual signals that are not elements of a subsequent action are tail wagging in dogs indicating a momentary mood (that we interpret as happy), or the tail between the legs indicating a subordinate social status.

Considering the various sensory modalities, visual displays are the most obvious to human observers and hardly need emphasizing. Visual cues are readily detected by people and the changes over time are observable, whereas other types of stimuli are less easily noticed and described. Thus, behavioral descriptions typically emphasize visual occurrences. In this book many examples of visual displays are integrated with the descriptions of social behavior. The discussions of threat behavior (in this

Figure 2-7 Visual displays in horses are especially obvious. The illustrations show: (top) a mild threat by the horse on the right to the horse on the left; (middle) a more extreme threat by the horse on the right to the horse on the left; and (bottom) signs of excitement indicated by dilation of nostrils compared to the rather noninvolved look on the left.

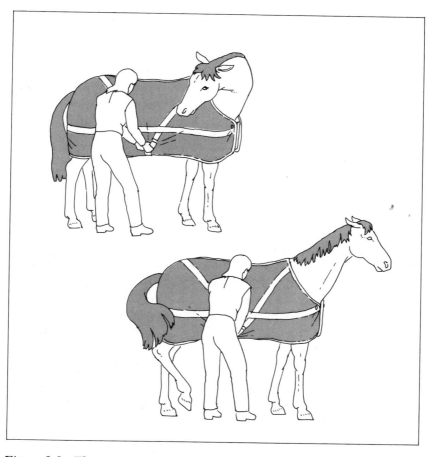

Figure 2-8 The same visual displays in horses can be directed toward other horses or people. Here we see increasing intensities of threat displays (from top to bottom) directed toward the handler. Note the tail swishing in the lower picture.

chapter, including Figures 2-2 through 2-5) and of sexual behavior provide numerous examples of postures that attract males and females to each other or otherwise give signals to conspecifics.

Physical Contact

This is the least used channel of communication. One example is the grooming of a dominant animal by a submissive individual to appease the dominant one and indicate subordi-

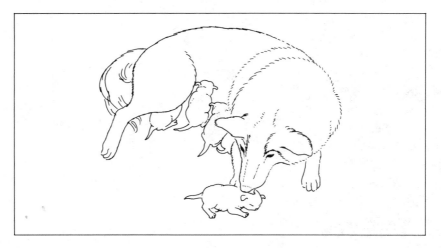

Figure 2-9 Physical contact by means of licking the pups is one way the mother dog has of communicating her location to the pups.

nation. This is common in primates. Another example is the inhibited ("soft") bite of dogs that is directed toward other dogs or people. Since the bite does relatively little physical harm, it represents a form of communication rather than an attack. The mother of a litter of pups that retrieves her newborn to the nest by licking them is using contact for communication since the young, which cannot hear vocalizations, can follow her wet tongue to her side (Figure 2-9).

Social Odors and Chemical Communication

Like ultrasonic emissions, chemical messages are usually inconspicuous to us. In humans the area of the olfactory epithelium of the nasal cavity is very limited compared with that of other animals, and even the parts of the brain devoted to processing olfactory information are very small. To reflect these anatomical differences, humans are referred to as "microsmatic" and domestic animals as "macrosmatic" (Figure 2-10).

An animal's olfactory world is completely different from ours. Many animals identify one another on the basis of odor, and their physiological states can be determined as well. Animals can smell where others have recently passed and how long ago they were there. This acute olfactory ability may help territorial animals in not getting lost because they can follow their own scent and even the direction of the scent trail by noting how old

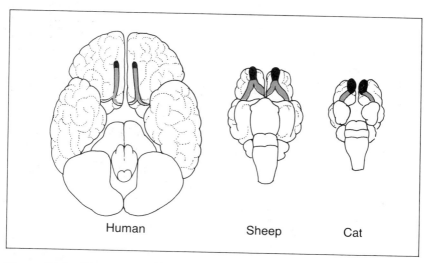

Figure 2-10   The greater olfactory ability of animals is evident at the anatomical level. Differences in the olfactory bulbs (solid black) and tracts (gray) are obvious in this ventral view of the human, sheep, and cat brains drawn to the same scale.

it is. One way dogs and cats may find their way around neighborhoods is by following odor gradients from factories, freeways, and the like.

Animals may also react to olfactory stimuli from other species. This is the way some prey animals are located. Predators may also be avoided by an animal using olfactory information. Rats, for example, exhibit a freezing behavior that interrupts feeding, sexual, and even maternal behavior when the odor of a cat is introduced (Griffith, 1920).

Aside from enjoying a much more vivid olfactory world than we do, other mammals experience a world of social odors that we are unable to appreciate. Social odoriferous secretions are produced in the urine, feces, saliva, and specialized skin glands.

Deliberately depositing odoriferous secretions in the environment is referred to as scent marking. There has probably been too much emphasis on the territorial function of scent marking and not enough on the role that the odors may have on an animal's emotional attachment to a marked location. This perspective of marking has been suggested by Ewer (1968) and Eisenberg and Kleiman (1972). An optimum odor field, according to Eisenberg and Kleiman, may provide an optimum level of security for each individual depending on age, sex, mood, and reproductive state. The odor field, consisting of olfactory stimuli from the individual animal, conspecifics, and the environment, may be disturbed through dissipation of the scent or the introduction of scents from other animals. Also a physiological change in the animal may alter its olfactory sensitivity and change the criteria as to what constitutes an optimum odor field.

We can see the concept of the odor field at work in male dogs and cats. A dog tends to freshen up its urine marks from time to time, but its urine marking is not noticeably frequent at home. Let a new dog in to mark the place up, or bring the old dog home after a prolonged absence, and you will see a flurry of urine-marking activity. This effect of intense marking in a new home followed by a tapering off is shown in Figure 2-11.

Tomcats also regularly engage in urine marking by spraying urine, and they frequently freshen up previously sprayed areas. Castrated male and spayed female cats generally do not engage in spraying, but they may spray as much around the house as a tomcat when they are emotionally disturbed by the introduction of female cats or a change in the household routine (Hart, 1980; Hart and Cooper, 1984).

The social odors produced by animals fall into two groups as outlined by Brown (1979) in his comprehensive review of social odors. Identifier odors are those produced by the normal metabolic processes and are stable for long periods of time. Emotive odors are those produced or released in special circumstances such as when an animal comes into rut, estrus, starts lactating, becomes stressed, or changes from a subordinate to a dominant position.

The identifier odors include those that are sex-specific (distinguish males from females), age-specific (distinguish animals of different ages), species-specific (characterize all members of a species), colony-specific (characteristic of all members of a nest, den, or group), and individual-specific (unique to an individual). There are numerous examples from the mammalian world of the ability of animals to make all these differentiations. Animals of every species tested so far are able to recognize different individuals on the basis of urine odors. This includes pigs, sheep, goats, and calves (Baldwin, 1977; Baldwin and Meese, 1977; Meese et al., 1975). The fact that human odors give us a chemical fingerprint or olfactory signature is evident by an animal's recognition of our odor. Kalmus (1955) found that dogs

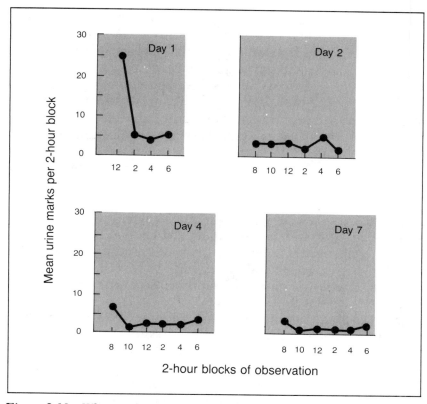

Figure 2-11 When male dogs are first put into a novel pen, they urine-mark very frequently. Later, presumably as the odor field becomes established, the frequency of urine marking declines to a rather infrequent baseline level. (From Hart, 1974*a*).

could discriminate between odors of unrelated people or members of the same family but had trouble distinguishing between identical twins.

The chemicals underlying social odors are almost entirely contained in urine or glandular secretions. These chemicals may be affected by internal metabolism, hormones, the chemistry of the glands involved, and the action of external microorganisms on these secretions.

Pheromones   Social odors that serve as a medium for chemical communication between conspecifics are referred to as pheromones. The most important characteristics of pheromones are their effectiveness over distance and the persistence of the signal over time, allowing communication in the absence of the signaling animal. Some of the best examples of the functions of pheromones come from studies of invertebrates. For example, the roosting aggregation of butterflies is a response to a chemical substance secreted by some animals. The swarming of bees at the entrance of a hive is facilitated by the secretion of a hormone by just one, or a few, worker bees. In other animals, a pheromone may act as an alarm signal. A vivid example is the chemical substance liberated into the water when a certain species of minnow is wounded; the substance, when detected by other animals of that species, results in rapid dispersal.

Pheromones are commonly used in insects for sexual attraction. A classic example is the chemical substance liberated into the air by the female silk moth, which can be detected in minute concentrations by males two or three miles away. By simply flying into the wind in response to the chemical signal, the male is able to come sufficiently close to the female to be able to visually locate her.

Among mammals in general, pheromones are secreted in the vaginal mucus and urine of females, presumably to communicate receptivity. Genital olfactory investigation, as well as investigation of female urine by males, is a reflection of the male's interest in the sex pheromones of females. Boars and males of some other wild species secrete a pheromone that is attractive to estrous females and seems to indicate to the female that the male is sexually mature. This pheromone enhances the receptive behavior of gilts and sows.

Flehmen is a behavior displayed frequently by bulls, rams, stallions, pigs, and even cats when they are allowed to investigate females or voided urine from females (Figure 2-12). The behavior is related to attempts by the male to utilize the vomeronasal organ and accessory olfactory system to examine the urine or genital secretions for the presence of nonvolatile sex pheromone (Estes, 1972; Hart, 1983; Ladewig and Hart, 1980; Ladewig et al., 1980). Although females occasionally display flehmen, the behavior occurs much more frequently in males and is therefore sexually dimorphic. Castration of male goats reduces the frequency of flehmen (Hart and Jones, 1975). The role of flehmen in sexual behavior will be dealt with in Chapter 3.

Scent Marking   The deliberate marking of environmental objects with urine, feces, saliva, or secretions from specialized skin

Figure 2-12 Flehmen is the term for the species-typical behavior illustrated above, which is involved in chemosensory investigation. The head is usually held horizontally and the upper lip retracted. This behavior shown here for the bull is involved in bringing materials into contact with the vomeronasal epithelium, which is part of the accessory olfactory system.

glands is referred to as scent marking. Sometimes this behavior obviously involves communication, as when an estrous female cat sprays vertical objects with her urine. Probably scent marking is more often related to the establishment of the odor field. There are some interesting behavioral patterns associated with scent marking, and I shall discuss some of these.

Feces are used in scent marking by various wild species. Many antelope species including gazelles, topi and hartebeest have specific dunging areas in their territories that the animals repeatedly freshen with the addition of new feces. The giant rat has the interesting behavior of climbing up a wall or another vertical object with its back legs and depositing fecal material above the ground, where it is apparently more effective (Ewer, 1968). Some of the scent-marking characteristics of feces undoubtedly reflect the secretion of glands into the feces. Although feces do not appear to be obviously used by most dogs in scent marking, they are used by wolves (Peters and Mech, 1975). Feces are sometimes left in conspicuous locations in a wolf's territory such as on rocks and trail junctions. An interesting variant of

Figure 2-13 The fecal scent marking being displayed by this male beagle dog is somewhat unusual. This particular dog would mark a tree with a handstand posture several consecutive times. It was conjectured that perhaps the behavior reflected an uncommon genetic link to wild carnivores that mark with feces. (From Hart, 1969).

fecal scent marking in a dog is shown in Figure 2-13. Perhaps this reflects an uncommon genetic link to the wild carnivores that mark with feces. Studies of feral horses show that males deposit dung heaps, and that fecal deposition is commonly associated with male-male interactions (Feist and McCullough, 1976; Salter and Hudson, 1982).

Urine is one of the most frequently used scent-marking substances of all mammals (Figure 2-14). The European lynx and wildcat bury their feces and urine (like the domestic cat) when they are well within their territorial limits. However, on the boundaries of their territories they have been observed to deposit feces and urine in rather conspicuous places.

Studies on the wolf indicate that our rather cursory interpretation of urine marking in the dog may be glossing over

Figure 2-14 Urine marking is common in male dogs and cats. Urine spraying by cats in the home is one of the commonest behavioral problems for cat owners.

much. Field observations show that wolves respond differently to marks from animals of different ages, and they can detect the time that has elapsed from when other animals were in the area. At the border of the territory they can tell approximately how long ago their neighbors passed through. Urine marks are more frequent during the breeding season of wolves, indicating a hormonal effect (Peters and Mech, 1975). In the wolf, urine marking occurs mostly at territorial boundaries. Interestingly, in wolves only breeding males appear to urine-mark. Nonbreeding adult males in a pack rarely do so. Lone wolves, which do not have territories and do not possess mates, rarely mark by the raised-leg urination (Rothman and Mech, 1979).

In dogs and cats urine marking is characteristic primarily of males and is reduced by castration. Urine marking by male dogs in the house can occur so frequently that they are castrated in an attempt to eliminate the behavior (Hopkins et al., 1976). In cats urine spraying is the most frequent behavioral problem for which veterinary treatment is sought.

A behavior often associated with urine marking and sometimes with defecation in dogs is ground scratching. The dog slashes the ground with one or more paws and usually leaves a visible mark (Bekoff, 1979a). The behavior also occurs in wolves (Peters and Mech, 1975). The notion that scratching helps disperse the scent seems untenable because the deposited urine or feces are rarely hit. Bekoff has postulated that scratching is a visual signal with the purpose leaving scratch marks on the soil. The behavior is also a visual display when other dogs see the scratching behavior. The latter view appears to be supported by Bekoff's observation that scratching was much more prominent when the scratcher had an audience of other dogs (Bekoff, 1979b). Scratching may also leave a chemical mark from secretions of the interdigital footpads. In this way scratching may be vewied as a composite signal consisting of both visual and chemical signals.

An interesting type of scent marking is found in domestic goats and some wild ungulates such as Barbary sheep, mountain goats, and bighorn sheep. Prior to and during the mating season males impregnate their fur with their own urine. Male goats urinate on the chest, neck, front legs, beard, and throat. Urination even into the mouth is common. The function of this behavior, referred to as self enurination or scent urination, has not been established experimentally. It has been suggested that the behavior functions to hasten or to synchronize the onset of estrus in females through olfactory stimulation (Coblentz, 1976).

It is also possible that the urine reveals the male ungulate's physiological condition to other males as indicated by metabolic by-products. For a dominant male his urine odor may reveal a superior state of nutritional well-being. By the same token the urine odor may also be a cue by which subordinates recognize a decline in condition of the dominant from aging or even a declining food intake.

Saliva is used by several species of marsupials for territorial identification. Some wild bears have been observed to scratch the trunks of trees and chew bark. The same animals have been observed to urinate on the tree, thus presumably duplicating their scent marking with another substance (Ewer, 1968). During mating, domestic pigs salivate excessively and an androgenic steroid in the saliva acts as a sex attractant to females (Mykytowycz and Goodrich, 1974).

Secretions of anal glands are used by a number of carnivores for scent-marking purposes. In some species the glands are simply pressed against an object. In others a type of anal drag is used in which an animal sits down so that the anal gland openings as well as the anus are resting on the ground and then pulls itself forward in this sitting posture with the front legs. Animals exhibiting this behavior are the anteater, badger, and polecat.

The scooting behavior of the domestic dog, which is practically identical to the anal drag just described, is usually attributed to discomfort resulting from impaction of the anal glands. There are some who believe that such scooting in dogs represents a type of scent marking. However, despite repeated attempts, no scent-marking or communicative functions have been associated with these glands (Doty and Dunbar, 1974).

The existence of scent glands in the skin is widespread among mammals, and scent marking with these glands is a common method of olfactory communication (Johnson, 1973; Muller-Schwarze, 1974; Thiessen and Rice, 1976). In many animals, specialized sebaceous skin glands are located around the head. In ground squirrels, porcupines, and other rodents, such glands are located at the angle of the mouth. A scent gland is found under the chin in rabbits.

A number of hoofed animals, including goats, deer, and antelope, have scent glands located at the inner corner of the eye or on the forehead near the case of the horn. These animals scent-mark with these glands by rubbing areas at the base of the horn or around the inner surface of the eye against tree branches or other objects.

Another group of scent glands is those found on different parts of the body that produce secretions that may be rubbed onto objects unintentionally during an animal's normal movements about its territory. For example, there is a scent gland on the top side of the tail of wolves (missing in the domestic dog) that is rubbed against overlying rocks, branches, or earth as the animal enters and leaves its den. Other glands serving a similar function are located on the flanks or in the pectoral region.

The domestic species in which scent marking with specialized glands is the most obvious is the cat (Hart, 1977a) (Figure 2-15). Tree (or furniture) scratching in cats creates a visual mark and could be called a visual display. When a cat is scratching a tree it is probably also applying secretory material from foot glands on the tree to give it a particular scent. This odor can then be detected by the resident and intruders, who would recognize this tree as some sort of territorial mark. Perhaps the sight of a scuffed-up tree attracts other cats to the visual mark to investigate it more fully.

Cats also engage in cheek or head rubbing. In these instances they presumably rub glands, located at the corners of the cheek or in the supraorbital region between the eye and the ear, on

Figure 2-15   Domestic cats display two types of scent marking with sebaceous skin glands. During scratching on a tree, secretions from foot glands are applied to the tree, leaving a chemical mark as well as a visual mark. Rubbing secretions from glands at the corner of the mouth is also commonly practiced by cats. These types of scent marking are not sexually dimorphic.

objects in the territory. Table legs, chairs, and doorway thresholds are common objects marked in this manner, as are the legs of the people in the cat's immediate family. This type of scent marking probably helps a cat to feel more familiar with its surroundings. Marking with glands of the feet or head is not sexually dimorphic in cats.

Substances used for scent marking on territorial boundaries appear to be regularly visited by neighboring animals and thus probably give animals the means of keeping track of their neighbors' presence without actually having to encounter the neighboring animals. The age of the substance could also convey information about a neighbor. For example, the absence over a long period of time of any freshly deposited material would perhaps indicate that a territory is open for expansion or occupancy by another animal. An unfamiliar substance would alert the resident animals to the presence of a new animal in the vicinity. Fresh urine or feces left as a scent mark serve as a warning to an animal following behind that the trail immediately ahead is occupied. Older scent marks would indicate that the trail is open (Ewer, 1968).

Comparison of Animal Communication and Human Language

Human language involves the use of words and sentences. Words are combined in a sentence with particular grammatical relations to produce a complete semantic proposition. The formation of sentences has traditionally been considered a uniquely human ability. Using a finite number of words, we can formulate an almost infinite number of meaningful sentences. We can say something that has not been said before. We can also refer to events or objects at a location far from ourselves or removed in time. In the context of a sentence, we can use a word that means one thing, but in another sentence use the word to mean something completely different.

It seems almost absurd to compare any type of animal communication with ours. However, there has been an attempt recently to see if the more intelligent nonhuman animals, namely chimpanzees and gorillas, can perform the most elementary functions of language. If so, then we can no longer claim exclusive domain of language.

Animal signals, such as alarm sounds or mating vocalizations, are usually genetically fixed. Notable exceptions are the dialects of birds and certain mammals observed in different environments. Our words, on the other hand, have arbitrary meanings that we have learned. There are probably no examples in nature of animals using purely learned sounds for communication. But in domestic animals we can teach simple words to pets. Hence dogs will learn to come at the call of their name and to sit when we say "sit."

Taking this use of language one step further, researchers in a number of laboratories have taught an impressive vocabulary of words to chimps. The Gardners taught their famous chimp, Washoe, signs of the American Sign Language (Gardner and Gardner, 1978). Premack taught the chimp Sarah simplified language using plastic chips (Premack, 1971). The animals can name objects and ask for things using the learned symbols. These and other studies are based on utilizing a visual medium to compensate for the ape's inability to articulate our kinds of sounds. It has been argued that chimps can formulate elementary sentences and therefore can use language (Patterson, 1978). On the other side of the debate are studies and arguments by psychologists, psycholinguists, and linguists claiming that an ape's language learning is extremely limited. Like dogs and cats, they learn isolated symbols but there is little evidence of mastery of elementary syntactic organization of language (Terrace et al., 1979). The argument may never be resolved and, for those outside the specialized field, the controversy is getting more esoteric. One notion is perhaps relevant in our interest in domestic animals: the extreme difficulty in showing that apes can communicate in a languagelike fashion points out the absurdity of the claims of some pet owners that they can "talk" to their dog or cat.

Communication and Deception

In this section I have stressed the importance of the use of signals to communicate information about the sender or for the sender to get a response from the receiver. The communication of mating readiness by a female can naturally benefit the sender and the male receiver. The same is true of agonistic social signals, mother-young interactions, and so forth.

Animal behaviorists are now recognizing the value of animals sometimes communicating misinformation, that is, being deceptive in their signaling. Consider the animal that has a less than optimal physical condition, through age or sickness, but maintains its superior dominance position with visual and auditory threatening signals. It deceives the receiver as to its real physical condition. Young animals in the process of being weaned display suckling activity with an intensity that is far out of proportion to their actual need. This can be deceptive to the mother, who is acting to reserve her resources for her next offspring.

In the sexual realm deception is more apparent. Animals appear not to have the counterpart of the classic human deception of the young man telling a young woman he is in love with her purely for the purpose of winning her affection for the night. In the animal world a sexual deception is practiced by females, however, in their secretion of sex pheromones at proestrus before coming into full estrus. Males are attracted to a female while she is still unwilling to mate. If more than one male is attracted, as is often the case, there will be competition between the males. By the time the female is ready to mate, she has the pleasure of mating with the strongest, fittest male. Thus this type of deception pays off for the female at the expense of the early bird males who catch her chemical signals but are beaten out by stronger males.

Although animal communication is nothing like human language, the more we learn about it, the more the complexities become apparent. Whether signals convey real information or misinformation is almost beside the point. Signals are used by one animal to manage or manipulate other animals in ways that have proved to be of selective advantage to the individuals through centuries of evolutionary adaptation.

■

AMICABLE AND AFFILIATIVE BEHAVIOR

Despite the fact that we make much of agonistic interactions between animals and the apparent vying for a position of dominance, most social animals seem to enjoy one another's company. They form special friendships, groom each other, and play together. Animals have ritualized greeting responses, usually involving mutual nose-to-nose or anogenital sniffing. These greeting responses function in the recognition and acceptance of group members.

Social Facilitation

A certain degree of social facilitation occurs in most animal groups. It is common to find in a group of cows or sheep almost all of the animals eating, grazing, or ruminating at the same time. The alarm bark of dogs or pigs triggers an alerting or orientation response on the part of all animals. Similarly, sheep tend to flock together under a threatening stimulus.

Another aspect of social facilitation is the tendency of one animal to follow another into a strange or threatening area. This is used by a horse handler to advantage when one wants to lead a group of horses through a narrow passage or over a bridge. One simply leads the animal that is the most easily coaxed into strange environments first and the others of the group will more readily follow.

Grooming

Social grooming is another type of affiliative behavior. Such grooming tends to be directed toward parts of the body that are

Figure 2-16 Amicable behavior occurs between a nomadic Fulani
herdsman and a cow as he strokes the perineal region. Another herdsman
strokes the animal under the jaw and neck. Cows appear to enjoy this
grooming and will sometimes even approach a herdsman as if to solicit it.
(From Lott and Hart, 1979.)

inaccessible to the animal being groomed, such as under the
chin, behind the ears, or over the base of the tail. The actual
amount of time devoted to social grooming is small. In a study
of dairy cattle, which socially groom by use of the tongue,
grooming constituted less than 1.5 minutes per day and occurred
during periods of resting. Generally the older cows did more
grooming and received more. Cows of higher social rank received more grooming than those of lower rank. Milk production
was correlated with the amount of grooming received. One investigator even goes so far as to suggest that if good groomers
in a herd are culled out, some degree of herd harmony may
suffer (Wood, 1977).

In cattle herds maintained by nomads, the herdsman can
take the role of another cow in social grooming; cattle seem to
accept this as they would grooming from a conspecific (Figure
2-16). One might classify the petting of dogs and cats by people
as a manifestation of grooming.

The advantage to the receiver of grooming is obvious. The
skin is conditioned, and parasites are removed. To the groomer,
perhaps salt from the skin is one source of motivation. The main

payoff to the groomer is probably in strengthening social and affiliative bonds, especially if the groomer is subordinate to the receiver.

Systematic observations on social grooming of other species are limited. Horses will at times stand adjacent to each other in opposite directions and groom the area over the base of the tail with the teeth. A licking and biting type of grooming occurs among canids, including the domestic dog. It is clear that pet cats and dogs usually enjoy being brushed and gently groomed by people.

Play

Play is a form of amicable behavior that commands considerable time from juveniles (Fagen, 1981). It is a regular part of the social behavior of all domestic mammals, but play is not universal among mammals. Small mammals that rapidly mature show little or no play behavior. Play is most prominent in young animals of those species with a relatively large brain and in which growth and maturation are relatively slow (Ewer, 1968).

What differentiates play from other types of social behavior such as serious fighting? Generally play behavior is not serious. Elements of aggressive behavior, killing of predators, escape from predators, and even sexual activity constitute play, but in play these elements occur in unpredictable rather than predictable sequences. Also, the elements of play may be repeated over and over and performed in an exaggerated manner.

In carnivores, which make much use of teeth, play is preceded by a signal indicating that "what follows is play." The most widely recognized example is the play invitational bow of canids (Figure 2-17). In young animals the bow rarely occurs outside the context of play. Interestingly, in adults the same display is seen as part of precopulatory sexual behavior, expecially by the female.

In dogs play bouts include wrestling, jaw wrestling, and inhibited biting. If there is an established dominance order, the dominant may allow a subordinate to act dominant. Mounting is a frequent aspect of play in male dogs. Male puppies mount other males as well as females. Mounting occurs as early as one month of age (Bekoff, 1972).

Cats are not social animals as adults, even though as kittens they interact and play socially. In cats play sequences are ini-

Figure 2-17   This play invitational bow of dogs is seen in several canid species.

tiated with a greater variety of postures. One is the belly-up posture, moving the back legs in a treading fashion. Two others are the pounce and sidestep in which a kitten walks sideways toward another. Mounting of females is not seen in cats until four months of age or later (West, 1974).

There are some common elements in canine or feline play. Both dogs and cats may play alone when partners are not available or interested in social play. In both species we see elements of prey capture and prey killing movements, with cats using the pounce and claws more and dogs favoring the biting and head-shaking style. Another theme in the play of both species is agonistic and fighting behavior similar to that used on conspecifics as adults.

In sheep and goats play behavior emphasizes running, jumping, and gamboling (jumping straight into the air). These behavior patterns are, of course, elements of predator escape or avoidance. Butting in sheep and rearing and butting in goats are aspects of agonistic behavior that constitute part of play (Sachs and Harris, 1978). Males engage in sexual play much more often than females, but females do display occasionally the male patterns of mounting and thrusting (Orgeur and Signoret, 1984). Sexual play is seen within the first week of age. In cattle, butting is common in both males and females, but pushing is more often engaged in by males. As with sheep, males also mount more often; this is usually directed toward other males. Flehmen is displayed by male calves, usually oriented toward females (Reinhardt et al., 1978).

In both herbivores and carnivores play behavior declines as the juvenile period of life phases into young adulthood. This is the time when prey capture or escape from predators becomes a matter of survival. It is also a time when agonistic behavior and sexual behavior take on adult seriousness.

An obvious benefit of play is to exercise the young. This is probably most important for nursing animals, which do not need to exercise to obtain food or shelter. Play has a function in facilitating the socialization of animals with other members of the same species. In fact, deprivation of play may lead to an impairment in normal social behavior (Fagen, 1981).

For all species, one of the most important functions of playing is the learning that occurs during play episodes. Play enriches the experiential world of the organism far beyond that which it would receive if it were only satisfying its immediate drives. The refinement of neuromuscular coordination involved in play hunting, fighting, and sexual behavior will be valuable to an animal when these patterns of behavior are needed in adulthood.

A number of ideas have been presented as to the evolution of play behavior. The functions or advantages of play such as exercise are obtained at some cost of energy, especially to the mother, who must still provide nutrition for the young. Therefore, some more long-term advantages of play in terms of inclusive fitness have been the subject of speculation (Fagen, 1976). Of particular interest is the notion that the games played are rehearsals for life-and-death situations in adulthood where there may be no opportunity to learn by mistakes or trial and error. The play of an ungulate, for example, includes much in the way of mock chase-and-escape activities. Learning to dodge and fake out the chaser helps a young ungulate develop the skills of escaping from a predator. Later in life escape behavior will have to be performed perfectly the first time it is needed if the ungulate is going to survive.

Chase and escape are also part of the play of young carnivores. Learning to anticipate quick turns and to calculate an effective diagonal approach to the animal being chased are skills for which development during play may be invaluable. Even the capture of a blowing leaf that bounces around helps a youngster develop skills at capturing something that twists and turns rather than moving straight out. To a carnivore, such refined capture skills may mean the difference between subsistence living and eating well enough to fight for sexual mates or raise healthy offspring.

In both herbivores and carnivores play fighting is common. However, play fighting is very different from actual fighting in both emotional effect and sequencing of behavior. But since the motor patterns are similar, this experience may give the players some edge in fighting as adults.

Since the development of predator escape skills, prey capture skills, or fighting skills necessitates the interaction of two or more young animals, it is necessary for the young animals to take turns in various roles. A dominant-subordinate relationship does not enter into the form of play.

The play in animals that we find such a delight is very serious business to the developing animal. Because the play behavior appeals to us, we have tended to breed for dogs and cats that play well into adulthood; this is termed neoteny. In the neotenized breeds, play behavior tends to characterize even the behavior of adults.

■

TYPES OF ANIMAL GROUPS AND TERRITORIALITY

Among wild species there is a great range in types of social organizations and territorial behavior. There are species in which an individual becomes the sole "owner" of various plots of land by defending its territory and other species in which individuals or groups of animals show practically no affinity for certain locations. Social and asocial species may or may not be territorial. For certain wild species territorial attachment may occur year-round or just during the breeding season. Sometimes within a territory there is a favored sheltered location that is defended, where the animal may have its nest or den. The animal may also have a home range where it freely roams for hunting or grazing without excluding others. Parts of the home range may be shared with other animals. Dogs and cats, as well as wild carnivores, usually have this type of territoriality.

We can assume that, as a result of social communication, animals are able to recognize others from neighboring territories and to know when their neighbors have disappeared or been replaced. Scent marking plays a role in establishing territorial boundaries and in influencing interactions among neighbors. Neighbor recognition occurs in solitary species as well as social species.

An important behavioral phenomenon occurs in relation to territoriality especially among fish and birds. Animals on their own territory display a sense of confidence and self-assurance. The same animals that have moved into the territory of another individual are fearful or apprehensive. In species that exhibit this type of territoriality the outcome of a fight between two animals is greatly influenced by its location. Even with considerable differences in size or age, there is a high probability that an animal will win a fight on its own territory and that it will lose a fight on another animal's territory. Domestic animals do not usually show this type of rigid territoriality. Dogs and cats, for example, while wandering through a neighborhood do not appear to be fearful or apprehensive when entering other yards (Leyhausen, 1965; Scott and Fuller, 1965).

Social-Territorial Arrangements of Wild Species

There are five general patterns of animal grouping and use of spaces. Some of these are organized around ownership of space or territory, and others form around social aggregations. When one of these systems applies to a domestic species (indicated below in parentheses), the reference is to the lifestyle of the wild ancestor or to animals of that species living freely in a feral situation.

Solitary-Territorial (Cats) In one type of arrangement, found in asocial species, a female and her young occupy a certain space and defend this space from other females and males of the same species. Males that are strangers are allowed into the female's territory only when she is sexually receptive. Males may also occupy a territory. Examples of species with this type of territorial arrangement are some squirrels, hamsters, shrews, and all feline species except lions and cheetahs. One feline, the feral domestic cat, is also included in this category, although there are many exceptions of multi-cat households where a number of adults seem to live on an amicable basis. Studies of feral housecats reveal they may occupy shared space at different times of day or share paths with cats in adjacent territories (Leyhausen, 1965).

Pairing-Territorial (*Songbirds*)   A second type of grouping, also seen in asocial species, is where a male and a female come together to share and defend a territory during the reproductive season. This is a common arrangement for backyard songbirds as well as for foxes and some other wild carnivores.

Female Grouping (*Cattle, Sheep, Goats*)   A third arrangement, found in some social species, is one in which a group of females with their young exist either free-ranging or on a home range. Males join the group just during the breeding season. When living apart from the females, the males may live singly or in all-male groups. A dominance hierarchy exists among the adult females and among the male groups. Species of deer, American bison, and elephants are examples of wild species existing in this arrangement. Cattle, sheep, and goats have ancestors that were characterized by this type of social pattern.

Harem (*Horses*)   A fourth grouping arrangement is that in which a single adult male lives together with a group of females, sometimes referred to as his harem. In some species the male defends a fixed site where the females reside, and in others he defends the moving harem. Males that do not have females may live in all-male (bachelor) herds. This arrangement may be permanent or seasonal and with differing degrees of territorial attachment. Such a pattern is found in impalas, lions, sea lions, and seals. In the case of sea lions, males arrive on a beach in a group prior to arrival of the females and set up small territorial sites that they defend. Later, when the females arrive, each male tries to gather together a sizable harem and attempts to retain his females within this territory. Some males attempt to steal from other harems. Young males coming into adulthood may attempt to stake out a territory and obtain a few females.

Herds of feral horses also have this social arrangement, but, since the group may often change location, defense of the harem is site-independent. One stallion customarily leads and defends a herd of mares and juvenile males and females. As the juvenile males reach adulthood the leader usually drives them off. However, some feral horse bands have been observed to include subordinate males that do no breeding (Berger, 1977; Miller and Denniston, 1979).

Recent observations of feral horses reveal that various bands of horses may interact, especially if they are forced together by limited vegetation or water supply. During a drought in Wyoming a study of several horse bands using a water hole in the Red Desert revealed that the bands interacted in a dominance fashion. The interband interactions involved not only the dominant stallion of each band but also the mares and even subordinate adult males that were part of the bands. Dominant bands had access to the water hole at will. Bands using the water hole often left the pool at the sight of a high-ranking band approaching. The highest-ranking bands never hesitated during their approach to the water, whereas the lowest-ranking bands sometimes waited longer than five hours to drink. The study included observations on 16 bands ranging in size from one to 16 horses. There was a strong correlation between harem size and the band dominance hierarchy, which in turn presumably reflected the ability of stallions to compete with other stallions for mares (Miller and Denniston, 1979).

Whereas the harem system is most characteristic of feral horses, other equids can live in a different life style. A study of feral asses in a desert region of California revealed that adult males lived solitary lives except when accompanying estrous females. The most stable social association was the mare-foal relationship. Females were usually seen alone with their foals or with other mares with foals. The investigator in this study felt that the loosely structured territorial system of the arid-land equids was adaptive to the low carrying capacity of the desert (Woodward, 1979).

Compound Social Structure (Dogs and Swine)   Finally there is the arrangement in which several adult males, adult females, and juveniles exist together as a group. The group may occupy a defended territory or a particular home range. In these situations there are separate dominance hierarchies for each group. Most males tend to be dominant over most females, and females are dominant over juveniles. This social-territorial arrangement is found in some large groups of wolves and is quite common among primates. Of the domestic species this grouping arrangement is characteristic of dogs and swine.

Dominance and Leadership Hierarchies

We have a tendency to believe that all groups of animals, whether they are formed under natural or artificial surroundings, establish a dominance hierarchy. There is no question that a peck order exists among some groups of mammals as it does among flocks of chickens. Such a social rank order is based on serial submissions in the groups such that animal A dominates B, who dominates C, who dominates D, and so forth. A, of course, dominates all of the animals, and B dominates all those except A.

As animals are studied more in the wild or feral state we are learning that hierarchies are not invariably present. In groups of mixed sexes and ages observed in the wild, interactions can be learned and based on roles rather than a drive to be dominant. Rowell (1974) notes that in monkeys, for which the dominance hierarchy concept seems so appropriate, social relationships can be explained in terms of fairly ordinary learning. In captive groups of monkeys Rowell describes hierarchies, but she notes that the hierarchy is more a function of some animals acting particularly submissive, probably from succumbing to stress-related diseases of captivity, than a drive of some animals to be dominant. The confident and cool nature we ascribe to the dominant monkey is more or less the rule of all group members in the wild.

Perhaps a fair analogy of the difficulties in understanding social order is to point to the fallacy of using the military rank concept to describe human social interactions in a mixed community. The local tennis champion may in some way be dominant at tennis club parties but socially quite subordinate at a company party. Our society, and those of animals, does not function with invariant lines of authority. We see in feral groups of domestic animals variations from the conventional social rank. Stallions, for example, are usually dominant to their mares and all juvenile horses. However, in observations of a herd of Camargue horses it was found that the herd stallion was subordinate to the mares, probably reflecting the fact that he was still a little immature and was a newcomer into an established group (Wells and Goldschmidt-Rothschild, 1979).

The fact that most herds or flocks of domestic animals are so uniform in size, age, and sex and have to compete for feeding space or rest areas promotes the organization of dominance hierarchies along the lines of a peck order. The above reservation notwithstanding, clear dominance hierarchies have been observed in chickens, pigs, cattle, sheep, goats, horses, and dogs.

Dominance Hierarchies The dominant-subordinate relationship may be of an absolute type, in which the subordinate never fights back, or it may be more relative, with one animal dominating the other most but not all of the time. The absolute dominant-subordinate relationship where the submissive animals do not hit back seems to characterize the peck order pattern seen in chickens, where higher-status hens also spend more time feeding (Tindell and Craig, 1959). The relative relationship has been observed in pigs (Beilharz and Cox, 1967). Small groups of growing pigs are usually organized into linear dominance hierarchies, but even the nondominant animals win some encounters. Recently, it has been suggested that the relative type of relationship is a characteristic of immature animals, and that it changes to an absolute type with adulthood (Beilharz and Zeeb, 1982). Parallel relationships, in which neither animal appears to be dominant, are also sometimes observed between individuals of several domestic species.

Nomadic herdsmen, who are with their cattle virtually all day, every day of the year, enter into the dominance hierarchy of their cattle by taking the role of absolute dominance (Lott and Hart, 1979).

Leadership In animal groups that have access to pastures, we see that some animals play the role of leaders in the movement from one location to another. An analysis of leadership in a flock of sheep revealed that being a leader is more a consequence of being somewhat a loner and independent in moving about (and followed by others) than in being part of a tightly knit group (Arnold, 1977). The leader in sheep and cattle herds is not usually the dominant animal. The leader is not always the same and the order of animals in following is not always the same. Just as a nomadic herdsman can be the dominant individual in his herd of cattle, so too can the herdsman take the role of leader and have his cattle follow behind (Figure 2-18) (Lott and Hart, 1979).

The first cow into the milking parlor is the same day after day but is not necessarily the dominant cow nor one of the lead cows (Kilgour and Scott, 1959; Soffie et al., 1976). She is, how-

Figure 2-18 A young Fulani herdsman briskly leads a herd of cattle to a new campsite. This illustrates the ability of a herdsman to take the role of the lead cow. (From Lott and Hart, 1979.)

ever, fairly high in social position. The entry order of the rest of the herd can vary some, although cows that tend to form pair bonds come into the parlor together (Gadbury, 1975). Cows yielding a higher amount of milk come earlier for milking and those yielding a lower amount of milk come later (Rathore, 1982). There appears to be no association between milk production and dominance rank. Interestingly, attempts to force a predetermined milking order upon a herd of cows were not successful (Albright et al., 1966).

Types of Dominance Hierarchies The extension of dominance-subordinate relationships throughout a group of animals results in a dominance or social hierarchy. The best known is the unidirectional or peck order relationship. The rigid unidirectional hierarchy is only one of several arrangements observed (Figure 2-19). The unidirectional hierarchy is found where animal A is dominant over B, B is dominant over C, C is dominant over D, and so forth. The unidirectional pattern is common in cattle herds and in most small groups of animals.

The triangular (multiangular) hierarchy is an arrangement probably found in all species, especially when the group size is large. Houpt et al. (1978) found that in small herds of horses the

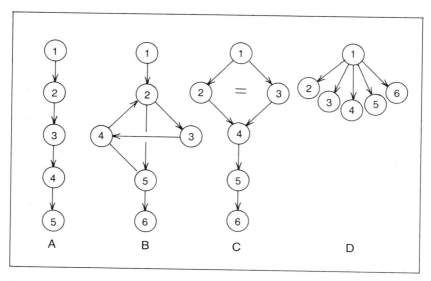

Figure 2-19 Schematic illustration of types of social hierarchies: A, unidirectional; B, C, triangular or parallel ranking within a unidirectional system; and D, a despotic system, in which one animal dominates all the others.

linear arrangement was formed but in large herds the triangular arrangement was common. In this arrangement animal A might be dominant over B, and B over C. However, C could be dominant over A. This arrangement often involves three animals of a larger group in which a basically unidirectional dominance hierarchy exists. Of course, in these cases the structure might eventually evolve to become more like a unidirectional hierarchy. Each hierarchy is a closed system, so that a new animal entering the group would assume the most subordinate position rather than a middle one.

Another type of social order exists in which one animal dominates all the others in the group and all subordinates appear to more or less share an equal rank. This despotic arrangement seems to characterize most groups of cats that are confined together (Hart, 1979).

Complex Hierarchies   Where there are large groups of animals consisting of several adult males, adult females, and juveniles, one is likely to find a separate social order for the females and juveniles. In the overall group, adult males tend to be dominant over all or most adult females and both males and adult females are dominant over all juveniles. One of the common examples of this complex social order is in large tribes of primates. In such primate groups it is not uncommon to find a female that is relatively high in the female dominance hierarchy and is dominant over some of the lower-ranking males.

The existence of a social organization provides order that is obviously highly adaptive to wild and domestic animals. The order allows each individual in the group to obtain food, rest, and other necessities that would not be possible without a social order because fighting and competition is reduced. The existence of a group also offers some protection from predators, care for the young, and efficient mating arrangements.

Although a group life offers advantages to both dominant and subordinate individuals, there are major advantages in being the most dominant. The dominant animal is able to feed and drink at will. Subordinates will often leave a food or water container on seeing a dominant animal approach. Here we see evidence of Rowell's contention that a dominance hierarchy might better be viewed as a subordinate hierarchy, because the social interactions are the function not so much of the behavior of the dominant animal but of the subordinate's behavior.

Dominant animals do sometimes exercise their "rights" to bully other animals at will. Cows of a high position in the social rank may drive subordinates away from a feed trough even though they appear not to be interested in eating themselves (Wagnon, 1965). In the wild this behavior has the advantage of conserving feed for the dominant individuals, but on the domestic scene it may deprive other animals of needed food. Increasing hunger may induce subordinates to become more resistant to yielding to dominant cows. The end result, however, can be undernourishment of subordinates to the point where there is substantial weight loss (Wagnon, 1965). Dominant cattle move into new grazing areas first, with lower-ranking animals behind. This gives the more dominant animals a better selection of foliage (Ewbank, 1969).

Dominant animals usually have the most frequent access to females as well. Just the presence of a dominant animal will often suppress mating attempts by subordinates. In a flock of sheep with several rams, ewes tend to mate with the dominant ram. If for some reason the dominant ram is infertile, then a number of ewes will not conceive and the flock fertility will be lowered

(Fowler and Jenkins, 1976). Even the sight of a dominant ram may be important. It has been noted that subordinate rams mounted and ejaculated less when viewed by dominant rams that could not even get to them than when tested alone (Lindsay et al., 1976). The practical consequence of this is that if a farmer wishes to use a subordinate ram as a stud, the breeding should be done some distance away from dominant males.

Because of the advantages that accrue to a subordinate reaching the dominant position, the dominant animal is always under some sort of social competition for its position.

Factors in Establishing and Maintaining Social Ranks

Social orders may be formed during the growth and maturation of young animals that are kept in the same location where they were born. Other social orders are formed when animals are grouped together for the first time. In domestic animals, social orders are frequently formed by a combination of these situations. Depending on the way the group has been formed, factors that influence the social rank of any animal may have differing degrees of importance. Individual recognition facilitates the stability of the group organization. Humans are usually in a dominant position over their animals as a function of several factors that are discussed in this section.

Size   This is an important and often determining factor of dominance. Social rank in pigs is positively correlated with body weight (McBride and Wyeth, 1964). Size was also found to be closely correlated with dominance in young beef cattle (Stricklin et al., 1980). In relationships between young and growing animals this factor is particularly important. The runt of a litter is almost always the most subordinate.

Presence of Weapons   The best example of the influence of weapons is the effect they have in cow herds. In one study it was found that the presence of horns was the major determinant of social position when the hierarchy was being established in a group of heifers for the first time. Body weight differences were of secondary importance (Bouissou, 1972). Wagnon (1965) noted in one herd that a cow used her six-inch horn stub (resulting from improper dehorning) to great advantage in achieving an admirable social status. When a hierarchy already exists, de-horning cows may not necessarily alter their social position.

Territoriality and Seniority   One factor that is particularly important in determining social rank relates to territoriality. When a group of animals is assembled, those that have previously lived in the area the longest and are the most familiar with it tend to be the most dominant. A somewhat similar concept holds for the role of seniority in that those animals who have been in a group the longest tend to be the most dominant. Humans often assume the dominant role over animals by virtue of raising them from infancy and hence being the most senior. By the same token, females are often dominant to their offspring even when the offspring reach or exceed their size. In a herd in which animals are moved in and out (for economic reasons) both seniority and size are correlated with age. Thus, older animals are often the most dominant. However, eventually aging takes its toll, and the oldest animal may decline in rank (Reinhardt and Reinhardt, 1975).

Temperament   An important factor in determining the outcome of encounters between animals is individual aggressiveness or pugnaciousness. Farmers sometimes observe that the most scrappy animals are the most dominant. Aggressive temperament has been related to dominance rank in horses (Houpt et al., 1978).

Some breeds of domestic animals are more aggressive than others and tend to be more dominant. For example, a large white breed of Australian pigs was found to be more aggressive than Berkshires (McBride and James, 1964). The same phenomenon exists among breeds of dogs, with terriers being notoriously aggressive and hence assuming a dominant position with respect to dogs from less pugnacious breeds.

Testosterone seems to increase an animal's ability to win encounters. This is true even when the hormone is given to females. In a study of dominance rank in cows, Bouissou (1979) found that testosterone injections resulted in heifers moving up in a dominance hierarchy. This is postulated to be due to a decreased fear of more dominant animals and suggests that a reduced tendency to take a submissive stand—which is different than an increased tendency to be aggressive—plays a role in determining dominance position.

Alliances and Nepotism Associating with dominant animals tends to increase another animal's social rank. This may be a function of friendship, and it is often seen when the offspring of a dominant female assumes a high rank by virtue of its association with the mother. This is common in cattle when a cow's female offspring remain as part of the herd. Such nepotism also occurs in horses (Houpt et al., 1978).

Sex In general males, even castrated males, are dominant over females.

Technology People can dominate animals stronger than they are. This is usually achieved by means of ropes, corrals, sticks, twitches, and choke chains. Were it not for our ability to use such tools, we would often be dominated by animals (Figure 2-20). Since we control an animal's access to food and water, this is another way of forcing our dominance. Demanding obedience and submission from dogs before giving them food and affection is an effective way of gaining dominance over them.

Formation of Animal Groups

Even in the most stable social orders there is a continuous testing and reinforcement of the dominance hierarchy by various types of threat behavior. Threats are frequently displayed when animals are fed or when a sexually receptive female is present. Generally the dominant animal threatens a subordinate one and the subordinate moves away or makes a submissive gesture.

In the domestic environment, animals are grouped and forced to interact with one another in ways that may be vastly different from the natural conditions to which their social behavioral patterns were adapted. Animal groupings are sometimes unstable, and new individuals may be added and others removed at frequent intervals. Competition for a high social rank may be more fierce than under natural situations because of excessive crowding and manipulation of the group composition. Males are usually castrated prepubertally to reduce intermale competition as well as to reduce sexual interest.

In the deliberate formation of animal groups on a farm or in a home, the goal should be to facilitate the rapid establishment

Figure 2-20 By the use of simple technology, such as a twitch used on a horse and a choke chain used on a dog, people are able to dominate animals more powerful than they. Interestingly, there is recent evidence that the twitch produces its calming and sedative effects by causing a release of endorphin (a naturally occurring morphinelike substance), rather than by producing moderate pain in the horse (Lagerweij et al., 1984).

(acceptance) of a dominance hierarchy. Sometimes this may mean helping a particular animal to be dominant. It is, of course, wise to help the animal that has factors such as size going for it already. A group might be assembled on this animal's territory. If outright fighting seems to be likely, regardless of territorial identity, it might be advisable to place all the animals that are to make up a new group in a new location. All of the animals will be slightly stressed or distracted by the new surroundings, which may allow stronger animals to become dominant with less fighting. Some animal handlers have suggested grouping animals in darkness.

The establishment of a dominance hierarchy with previously strange animals can involve considerable fighting. In pigs the fighting often lasts no more than 24 hours, and a hierarchy can be determined shortly thereafter. Most fighting is between animals that will take adjacent ranks (Meese and Ewbank, 1973). Even under the best management procedures, grouping animals may result in severe fighting. One approach that has worked in reducing fighting in pigs is to administer a tranquilizer to all animals. This seems to inhibit severe fighting but still allows for the emergence of a new dominance hierarchy (Symoens and van den Brande, 1969).

Individual or group recognition is a prerequisite for establishing and maintaining a dominance hierarchy. Studies have shown that animals probably use more than one sensory system in this recognition. For example, sight is not essential but is useful. Pigs that were made temporarily blind with contact lens could still form a dominance hierarchy when mixed with previously unacquainted blinded pigs. When the olfactory sense was also blocked, the formation of a dominance hierarchy was prevented (Ewbank and Meese, 1974). Visual recognition is also not necessary for individual recognition and maintenance of rank order with cows (Bouissou, 1971), but it is with poultry (Guhl and Ortman, 1953).

Animals may be removed from a group, isolated, and returned without being attacked, but the duration of time they may be removed appears to vary with social rank. One study found that top-ranking pigs could be safely returned after 25 days of isolation, whereas low-ranking pigs were severely attacked after three days of isolation. In some cases the returned top-ranking pig received attention in which it was nosed and licked around the head and anogenital region. When low-ranking pigs were returned, they were attacked most often by other low-ranking pigs, which would sometimes interrupt eating or another activity to attack them. If they remained outside the lying area, they were prone to fewer attacks (Ewbank and Meese, 1971). Eventually low-ranking animals that had been temporarily removed from the group were tolerated, and a day or so later they were allowed to eat with the group without being attacked.

In this chapter we have examined the ways in which animals interact socially with one another. Elements of social behavior involved in dominance hierarchies, aggressive interactions, and affiliative behavior have a genetic basis that relates to the species-specific lifestyle of the wild ancestors of our domestic species. Since the husbandry of domestic animals necessarily requires grouping several animals together, it is imperative that we understand and continue to examine animal social behavior to make our care effective and humane for both animals and people.

References

ALBRIGHT, J. L., W. P. GORDON, W. C. BLOCK, J. P. DEITRICH, W. W. SNYDER, AND C. E. MEADOWS. 1966. Behavioral responses of cows to auditory training. *J. Dairy Sci.* 49:104–106.

ARDREY, R. 1966. *The territorial imperative.* New York: Atheneum.

ARNOLD, G. W. 1977. An analysis of spatial leadership in a small field in a small flock of sheep. *Appl. Anim. Ethol.* 3:263–270.

ARNOLD, G. W., AND P. J. PAHL. 1974. Some aspects of social behaviour in domestic sheep. *Anim. Behav.* 22:595–600.

BALDWIN, B. A. 1977. Ability of goats and calves to distinguish between conspecific urine samples using olfaction. *Appl. Anim. Ethol.* 3:145–150.

BALDWIN, B. A., AND G. B. MEESE. 1977. The ability of sheep to distinguish between conspecifics by means of olfaction. *Physiol. Behav.* 18:803–808.

BEILHARZ, R. G., AND D. F. COX. 1967. Social dominance in swine. *Anim. Behav.* 15:117–122.

BEILHARZ, R. G., AND K. ZEEB. 1982. Social dominance in dairy cattle. *Appl. Anim. Ethol.* 8:79–97.

BEKOFF, M. 1972. The development of social interaction, play, and metacommunication in mammals: An ethological perspective. *Quart. Rev. Biol.* 47:412–434.

BEKOFF, M. 1977. Social communication in canids: Evidence for the evolution of a stereotyped mammalian display. *Science* 197:1097–1099.

BEKOFF, M. 1979a. Scent marking by free ranging domestic dogs: Olfactory and visual components. *Biol. Behav.* 4:123–129.

BEKOFF, M. 1979b. Ground scratching by male domestic dogs: A composite signal. *J. Mammal.* 60:847–849.

BERGER, J. 1977. Organizational systems and dominance in female horses in the Grand Canyon. *Behav. Ecol. Sociobiol.* 2:131–146.

BLEICHER, N. 1963. Physical and behavioral analysis of dog vocalizations. *Amer. J. Vet. Res.* 24:415–427.

BOUISSOU, M. 1971. Effect of precluding sight or physical contact on rank order behaviour in domestic cattle. *Ann. Biol. Anim. Biochem. Biophys.* 11:191–198.

BOUISSOU, M. 1972. Influence of body weight and presence of horns on social rank in domestic cattle. *Anim. Behav.* 20:474–477.

BOUISSOU, M. 1979. Effect of injections of testosterone propionate on dominance relationships in a group of cows. *Horm. Behav.* 11:388–400.

BROWN, A. M., AND J. D. PYE. 1975. Auditory sensitivity at high frequencies in mammals. *Adv. Comp. Physiol. Biochem.* 6:1–73.

BROWN, K. A., J. S. BUCHWALD, J. R. JOHNSON, AND D. J. MIKOLICH. 1978. Vocalization in the cat and kitten. *Devel. Psychobiol.* 11:559–570.

BROWN, R. E. 1979. Mammalian social odors: A critical review. In *Advances in the study of behavior*, vol. 10, pp. 103–162, eds. J. S. Rosenblatt, R. A. Hinde, E. Shaw, and C. G. Beer. New York: Academic.

CAIRNS, R. B., AND D. L. JOHNSON. 1965. The development of interspecies social attachments. *Psychonom. Sci.* 2:337–338.

COBLENTZ, B. 1976. Functions of scent-urination in ungulates with special reference to feral goats (*Capra hircus* L.). *Amer. Natur.* 11:549–557.

DE VUYST, A., G. THINÈS, L. HENRIET, AND M. SOFFÌE. 1964. Influence of auditory stimulations on the sexual behavior of the bull. *Experientia* 209:648–650.

DOTY, R. L., AND I. F. DUNBAR. 1974. Attraction of beagles to conspecific urine, vaginal and anal sac secretion odors. *Physiol. Behav.* 12:825–833.

EISENBERG, J. F., AND D. G. KLEIMAN. 1972. Olfactory communication in mammals. *Ann. Rev. Ecol. Syst.* 3:1–22.

ESTES, R. D. 1972. The role of the vomeronasal organ in mammalian reproduction. *Mammalia* 36:315–341.

EWBANK, R. 1969. Social behaviour and intensive animal production. *Vet. Record* 85:183–186.

EWBANK, R., AND M. J. BRYANT. 1972. Aggressive behaviour amongst groups of domesticated pigs kept at various stocking rates. *Anim. Behav.* 20:21–28.

EWBANK, R., AND G. B. MEESE. 1971. Aggressive behaviour in groups of domesticated pigs upon removal and return of individuals. *Anim. Prod.* 13:685–693.

EWBANK, R., AND G. B. MEESE. 1974. Individual recognition and the dominance hierarchy in the domesticated pig. The role of sight. *Anim. Behav.* 22:473–480.

EWER, R. F. 1968. *Ethology of mammals.* New York: Plenum.

FAGEN, R. M. 1976. Exercise, play and physical training in animals. In *Perspectives in ethology,* vol. 2, pp. 189–219, eds. P. P. G. Bateson and P. H. Klopfer. New York: Plenum.

FAGEN, R. M. 1981. *Animal play behavior.* New York: Oxford University Press.

FEIST, J. D., AND D. R. MC CULLOUGH. 1976. Behavior patterns and communication in feral horses. *Z. Tierpsychol.* 41:337–371.

FLYNN, J. P. 1967. The neural basis of aggression in cats. In *Neurophysiology and emotion,* pp. 40–60, ed. B. Glass. New York: Rockfeller University.

FOWLER, D. G., AND L. A. JENKINS. 1976. The effects of dominance and infertility of rams in reproductive performance. *Appl. Anim. Ethol.* 2:327–337.

FRASER, A. F. 1957. The state of flight or flight in the bull. *Brit. J. Anim. Behav.* 5:48–49.

GADBURY, J. C. 1975. Some preliminary observations on the order of entry of cows into herringbone parlours. *Appl. Anim. Ethol.* 1:275–281.

GARDNER, R. A., AND B. T. GARDNER. 1978. Comparative psychology and language acquisition. *Ann. N.Y. Acad. Sci.* 309:37–76.

GEIST, V. 1965. The evolution of horn-like organs. *Behaviour* 27:175–213.

GEYER, L. A. 1979. Olfactory and thermal influences on ultrasonic vocalization during development in rodents. *Amer. Zool.* 19:420–431.

GEYER, L. A., AND R. J. BARFIELD, 1980. Regulation of social contact by the female rat during the postejaculatory interval. *Anim. Learn. Behav.,* 8:679–685.

GEYER, L. A., T. K. MC INTOSH, AND R. J. BARFIELD. 1978. Effects of ultrasonic vocalizations and male's urine on female readiness to mate. *J. Comp. Physiol. Psychol.* 92:457–462.

GRIFFITH, C. R. 1920. The behavior of white rats in the presence of cats. *Psychobiology* 2:19–28.

GUHL, A. M., AND L. L. ORTMAN. 1953. Visual patterns in the recognition of individuals among chickens. *Condor* 55:287–298.

HAFEZ, E. S. E., AND J. P. SIGNORET. 1969. The behavior of swine. In *The behavior of domestic animals*, pp. 349–390, ed. E. S. E. Hafez. Baltimore: Williams & Wilkins.

HARRINGTON, F. H., AND L. D. MECH. 1978. Howling at two Minnesota wolf pack summer homesites. *Canadian J. Zool.* 56:2024–2028.

HART, B. L. 1969. Unusual defecation behavior in the male dog. *Commun. Behav. Biol.* 4:237–238.

HART, B. L. 1974*a*. Environmental and hormonal influences on urine marking behavior in the adult male dog. *Behav. Biol.* 11:167–176.

HART, B. L. 1974*b*. Gonadal androgen and sociosexual behavior of male mammals: A comparative analysis. *Psychol. Bull.* 81:383–400.

HART, B. L. 1977*a*. Olfaction and feline behavior. *Feline Pract.* 7(5):8–10.

HART, B. L. 1977*b*. Three disturbing behavioral disorders in dogs: Idiopathic viciousness, hyperkinesis and flank sucking. *Canine Pract.* 4(6):10–14.

HART, B. L. 1979. Feline life-styles: Solitary versus communal living. *Feline Pract.* 9(6):10–15.

HART, B. L. 1980. Objectionable urine spraying and urine marking behavior in cats: Evaluation of progestin treatment in gonadectomized males and females. *J. Amer. Vet. Med. Assn.* 77:529–533.

HART, B. L. 1983. Flehmen behavior and vomeronasal organ function. In *Chemical signals*, pp. 87–103, eds. D. Muller-Schwarze and R. M. Silverstein. New York: Plenum.

HART, B. L., AND R. E. BARRETT. 1973. Effects of castration on fighting, roaming and urine spraying in adult male cats. *J. Amer. Vet. Med. Assn.* 163:290–292.

HART, B. L, AND L. J. COOPER. 1984. Factors relating to urine spraying and fighting in prepubertally gonadectomized male and female cats. *J. Amer. Vet. Med. Assn.* 184:1255–1258.

HART, B. L., AND T. O. A. C. JONES. 1975. Effects of castration on sexual behavior of tropical male goats. *Horm. Behav.* 6:247–258.

HINDE, R. 1969. The bases of aggression in animals. *J. Psychol. Res.* 13:213–219.

HOPKINS, S. G., T. A. SCHUBERT, AND B. L. HART. 1976. Castration of adult male dogs: Effects on roaming, aggression, urine marking, and mounting. *J. Amer. Vet. Med. Assn.* 168:1108–1110.

HOUPT, K. A., K. LAW, AND V. MARTINISI. 1978. Dominance hierarchies in domestic horses. *Appl. Anim. Ethol.* 4:273–283.

HUSTON, J. F. 1978. Forage utilization and nutrient requirements of the goat. *J. Dairy Sci.* 61:988–993.

JOHNSON, R. P. 1973. Scent marking in mammals. *Anim. Behav.* 21:521–535.

KALMUS, H. 1955. The discrimination by the nose of the dog of individual human odors and in particular the odors of twins. *Brit. J. Anim. Behav.* 3:25–31.

KILEY, M. 1972. The vocalizations of ungulates, their causation and function. *Z. Tierpsychol.* 31:171–222.

KILGOUR, R., AND T. H. SCOTT. 1959. Leadership in a herd of dairy cows. *Proc. N. Z. Soc. Anim. Prod.* 19:36–43.

LADEWIG, J., AND B. L. HART. 1980. Flehmen and vomeronasal organ function in male goats. *Physiol. Behav.* 24:1067–1071.

LADEWIG, J., E. O. PRICE, AND B. L. HART. 1980. Flehmen in male goats: Role in sexual behavior. *Behav. Neural Biol.* 30:312–322.

LAGERWEIJ, E., P. C. NELIS, V. M. WIEGANT, AND J. M. VAN REE. 1984. The twitch in horses: A variant of acupuncture. *Science* 225:1172–1174.

LEHNER, P. N. 1978. Coyote vocalizations: A lexicon and comparison with other canids. *Anim. Behav.* 26:712–722.

LEYHAUSEN, P. 1965. The communal organization of solitary mammals. *Symp. Zool. Soc. Lond.* 14:249–263.

LEYHAUSEN, P. 1979. *Cat behavior: The predatory and social behavior of domestic and wild cats.* New York: Garland.

LINDSAY, D. R., D. G. DUNSMORE, J. P. WILLIAMS, AND G. J. SYME. 1976. Audience effect on the mating behaviour of rams. *Anim. Behav.* 24:818–821.

LORENZ, K. 1966. *On aggression.* New York: Harcourt, Brace & World.

LOTT, D. F., AND B. L. HART. 1977. Aggressive domination of cattle by Fulani herdsman and its relation to aggression in Fulani culture and personality. *Ethos* 5:174–186.

LOTT, D. F., AND B. L. HART. 1979. Applied ethology in a nomadic cattle culture. *Appl. Anim. Ethol.* 5:309–319.

MARLER, P. 1967. Animal communication signals. *Science* 157:769–774.

MC BRIDE, G., AND J. W. JAMES. 1964. Social behaviour of domestic animals. IV. Growing pigs. *Anim. Prod.* 6:129–139.

MC BRIDE, G., AND G. S. F. WYETH. 1964. Social behaviour of domestic animals. VI. A note on some characteristics of "runts" in pigs. *Anim. Prod.* 6:249–252.

MECH, L. D. 1970. *The wolf.* Garden City, N.Y.: Natural History.

MEESE, G. B., D. J. CONNER, AND B. A. BALDWIN. 1975. Ability of the pig to distinguish between conspecific urine samples using olfaction. *Physiol. Behav.* 15:121–125.

MEESE, G., AND R. EWBANK. 1973. The establishment and nature of the dominance hierarchy in the domesticated pig. *Anim. Behav.* 21:326–334.

MILLER, R., AND R. H. DENNISTON. 1979. Interband dominance in feral horses. *Z. Tierpsychol.* 51:41–47.

MOYER, K. E. 1968. Kinds of aggression and their physiological bases. *Commun. Behav. Biol.* 2:65–87.

MULLER-SCHWARZE, D. 1974. Olfactory recognition of species, groups, individuals and physiological states among mammals. In *Pheromones*, pp. 316–336, ed. M. C. Birch. New York: Elsevier.

MYKYTOWYCZ, R., AND B. GOODRICH. 1974. Skin glands as organs of communication in mammals. *J. Invest. Dermatol.* 62:124–131.

ORGEUR, P., AND J. P. SIGNORET. 1984. Sexual play and its functional significance in the domestic sheep (*Ovis aries* L.) *Physiol. Behav.* 33:111–118.

PATTERSON, F. G. 1978. The gestures of a gorilla: Language acquisition in another pongid. *Brain Lang.* 5:72–97.

PETERS, R. P., AND L. D. MECH. 1975. Scent-making in wolves. *Amer. Scientist* 63:628–637.

POINDRON, P., AND M. CARRICK. 1976. Hearing recognition of the lamb by its mother. *Anim. Behav.* 24:600–602.

PREMACK, D. 1971. Language in a chimpanzee? *Science* 172:808–822.

PRICE, E. O., AND J. THOS. 1980. Behavioral responses to short term social isolation in sheep and goats. *Appl. Anim. Ethol.* 6:331–339.

RATHORE, A. K. 1982. Order of cow entry at milking and its relationships with milk yield and consistency of the order. *Appl. Anim. Ethol.* 8:45–52.

REINHARDT, V., F. M. MUTISO, AND A. REINHARDT. 1978. Social behavior and social relationships between female and male prepubertal bovine calves (*Bos indicus*). *Appl. Anim. Ethol.* 4:43–54.

REINHARDT, V., AND A. REINHARDT. 1975. Dynamics of social hierarchy in a dairy herd. *Zeit. Tierpsychol.* 38:315–323.

ROTHMAN, R. J., AND D. L. MECH. 1979. Scent-marking in lone wolves and newly formed pairs. *Anim. Behav.* 27:750–760.

ROWELL, T. E. 1974. The concept of social dominance. *Behav. Biol.* 11:131–154.

SACHS, B. D., AND V. S. HARRIS. 1978. Sex differences and developmental changes in selected juvenile activities (play) of domestic lambs. *Anim. Behav.* 26:678–684.

SALES, G., AND D. PYE. 1974. *Ultrasonic communication by animals.* London: Chapman & Hall.

SALTER, R. E., AND R. J. HUDSON. 1982. Social organization of feral horses in western Canada. *Appl. Anim. Ethol.* 8:207–223.

SCOTT, J. P. 1948. Dominance and the frustration–aggression hypothesis. *Physiol. Zool.* 21:31–39.

SCOTT, J. P., AND J. L. FULLER. 1965. *Genetics and the social behavior of the dog.* Chicago: University of Chicago Press.

SEBEOK, T. A., ed. 1977. *How animals communicate.* Bloomington, Ind.: Indiana University Press.

SIGNORET, J. P., F. DU MESNIL DU BUISSON, AND R.-G. BUSNEL. 1960. Rôle dún signal acoustique de verrat dans le comportement réactionnel de la truie en oestrus. *C. r. hebd. Séanc. Acad. Sci. Paris* 250:1355–1357.

SOFFIE, M., G. THINES, AND G. DE MARNEFFE. 1976. Relation between milking order and dominance values in a group of dairy cows. *Appl. Anim. Ethol.* 2:271–276.

STRICKLIN, W. R., H. B. GRAVES, L. L. WILSON, AND R. K. SINGH. 1980. Social organization among young beef cattle in confinement. *Appl. Anim. Ethol.* 6:211–219.

SYMOENS, J., AND M. VAN DEN BRANDE. 1969. Prevention and cure of aggressiveness in pigs using the sedative azaperone. *Vet. Record* 85:64–67.

TERRACE, H. S., L. A. PETITTO, R. J. SANDERS, AND T. G. BEVER. 1979. Can an ape create a sentence? *Science* 206:891–902.

THIESSEN, D., AND M. RICE. 1976. Mammalian scent gland marking and social behavior. *Psychol. Bull.* 83:505–539.

TINDELL, D., AND J. V. CRAIG. 1959. Effects of social competition on laying house performance in the chicken. *Poult. Sci.* 38:95–105.

WAGNON, K. 1965. Social dominance in range cows and its effect on supplemental feeding. *Calif. Ag. Exp. St. Bull.* 819:1–32

WELLS, S. M. AND B. V. VON GOLDSCHMIDT-ROTHSCHILD. 1979. Social behaviour and relationships in a herd of Camargue horses. *Z. Tierpsychol.* 49:363–380.

WEST, M. 1974. Social play in the domestic cat. *Amer. Zool.* 14:427–436.

WOOD, M. T. 1977. Social grooming patterns in two herds of monozygotic twin dairy cows. *Anim. Behav.* 25:635–642.

WOODWARD, S. L. 1979. Social system of feral asses (*Equus animun*). Z. *Tierpsychol.* 49:304–316.

CHAPTER THREE ▪ SEXUAL BEHAVIOR OF FEMALES AND MALES

An understanding of various aspects of sexual behavior is of obvious importance in raising livestock species. Sheer reproductive success in terms of numbers of healthy offspring produced per year is of prime importance in sheep, swine, and beef cattle production. If we miss the behavioral signs of estrus in our plans to bring males and females together for breeding, reproductive success is markedly affected. In dairy cattle the number of offspring is less important, but the timing of births is critical, and a missed estrus period delays conception for almost one month with a significant effect on cumulative milk production. In the companion animal species there is less interest in numbers of offspring and more concern for producing particularly valuable offspring. Being able to differentiate normal from abnormal estrus and resolving problems with sexual inactivity can be critical in valuable animals. With the advent of artificial insemination, there is less attention devoted to the details of sexual interactions between males and females. However, even when semen is collected artificially, a familiarity with sexual behavior under more natural conditions is important. Understanding normal sexual behavior allows one to diagnose problems and to decide when the use of artificial insemination may be advisable. In pet animals and livestock raised by hand, undesirable aspects of sexual behavior, particularly mounting of people, may become a problem and warrant attention.

■

SEXUAL BEHAVIOR OF WILD ANCESTORS AND RELATIVES OF DOMESTIC SPECIES

To understand what is normal sexual behavior and what are behavioral changes induced by the process of domestication, it is important to know something about breeding strategies, seasonality, and polygamous versus monogamous pairing of animals in the wild state. Since all the domestic species are descendents of wild ancestors, we can use the wild ancestors or close relatives that live in the wild as points of reference. Keep in mind that under domestic conditions most domestic species do not display the full range of sexual behavioral patterns. Frequently, especially in large animal operations, the female is confined and the male brought to her stall or pen for breeding. Stud males may be so well adapted to a breeding regimen that they copulate with virtually no courtship or precopulatory behavior. The impact of farm management practices on the development and expression of sexual behavior in farm animals has recently been reviewed by Price (1984).

The closest example in livestock animals to a natural mating system is in sheep that are maintained in large pastures. One or more rams may be run with a number of ewes. Here we can see the interactions of social structure and sexual responsiveness. If more than one ram is present, a male social hierarchy is formed. It has been reported that the dominant ram may do most or all of the mating (Hulet et al., 1962). This is close to what would happen in the wild state. Knowing something about the dominance hierarchy is important because if the dominant ram happens to be less fertile or genetically inferior to the other rams, the productivity of the whole flock will suffer.

Breeding Strategies

In all species reproductive behavior reflects breeding strategies, which are usually concerned with leaving as many surviving offspring to reproduce as possible. The number of off-spring produced by an individual reflects his or her reproductive success or individual fitness. For the most part, males have the capability to, and an interest in, breeding several females. There is competition for breeding rights between males. The breeding strategy in a female is generally that of ensuring that as many of her offspring survive as possible. A female invests considerable time in caring for her offspring. The fact that swine, cats, and dogs give birth to several young, and ruminants and horses typically have only one or two young, is a reflection of the differential rates of infant and adult mortality and longevity of the wild ancestors.

Females of most domestic species ovulate spontaneously. In the cat and rabbit, however, ovulation is induced by copulation. Coitus-induced ovulation increases a female's reproductive efficiency, especially if males are not constantly around. Even in the spontaneous ovulators, coitus can facilitate or hasten ovulation. This has been observed in cattle, sheeps, and pigs (Jöchle, 1975). This consideration may be important in the utilization of artificial insemination. The effectiveness of artificial insemination in cattle is increased, for example, if the clitoris is massaged when the insemination is done (Randel et al., 1975).

One aspect of reproductive strategy involves efforts in helping the reproductive success of relatives (Hamilton, 1971). For example, a young stallion who is unable to maintain his own herd can increase his own genetic representation in future populations by supporting and defending a lead stallion that is related to him. This indirect, better-than-nothing strategy increases an animal's inclusive reproductive fitness. Females may also rely on the indirect route to genetic immortality by helping sisters when they themselves are not able to breed. This commonly happens in wolf packs: females that are usually related to the dominant female help her, and she is the only one to have pups.

Seasonality

Although most domestic animals were originally derived from wild species of the Middle East that were not markedly seasonal in reproduction, the current domestic species, adapted to temperate zones, stem from ancestors that were at one time seasonal in temperate zones. Most wild mammals of temperate zones are highly seasonal. The bearing and caring for offspring is extremely taxing on females. In the wild, selective forces result in a pattern of reproductive hormone secretion that restricts the birth of young to the most optimal time of the year. In the prey species concentrating births into one short time span also reduces the risk of predation on the offspring of any one female. Females of all species actively reject sexual advances of males outside the breeding season. Thus, copulation is closely timed to allow breeding when the probability of conception is highest and to enhance the chances that a female's ova will be fertilized by the strongest (most fit) male available.

Artificial selection over centuries of domestication has virtually eliminated reproductive seasonality in cattle, swine, and dogs; females of these species will give birth at almost any time of the year. In horses, goats, and sheep, seasonal influences exist but have been considerably weakened. The ancient Soay breed of sheep, which is native to Scotland, shows much more pronounced seasonal influences than the commoner breeds (Grubb and Jewell, 1973).

Males of some domestic species are also seasonal in their reproductive function. In horses, sheep, and goats, species in which seasonality is still evident among females, assays for blood testosterone concentrations in males reveal seasonal fluctuations. In addition to activating sexual behavior, of course, testosterone facilitates the display of other male behavioral patterns, particularly aggressive behavior. During the mating season male–male competition and fighting, which are facilitated by testosterone, are highly adaptable in that they give some males privileged access to females. During the time of year when females are not receptive, it would be of distinct disadvantage to males to be drawn into fights with other males and risk injury; hence the decline of testosterone, with the concomitant decline in male aggressiveness, is adaptive for the off-season.

Although the female dog is seasonal in the sense of going through two breeding periods per year at roughly six-month intervals, there is no tendency for this to occur at any particular time of year. Thus, male dogs must be sexually active throughout the year (Stebenfeldt and Shille, 1977). In contrast, there are clear indications of annual cyclicity in male gonadal function in wild canid species in which females are definitely seasonal.

Polygamous versus Monogamous Matings

Although the idea of monogamous pairings of animals has been highly popularized, particularly for birds, where more than 90 percent of all species are considered monogamous, less than 3 percent of all mammalian species fall into this category (Kleiman, 1977). The dog is the one domestic animal that had monogamous wild ancestors.

Wild species that have been selected for domestication as farm animals are polygamous. This behavior is advantageous on the farm, where husbandry has always been characterized by keeping one or just a few males around to service a relatively large number of females. Therefore, sexual promiscuity on the part of males and females, with infant care restricted to the female, has been a desirable mode of reproduction.

In the domestic dog behavior related to monogamous pairing has been selected against, but dogs still show signs of monogamous tendencies. The occurrence of mate preferences in dogs is well documented (Beach and LeBoeuf, 1967) and is probably more evident in dogs than other species.

There are other characteristics besides pair bondings that distinguish monogamous species. There is less sexual dimorphism, both behaviorally and morphologically. This, too, is evident in the dog: differences between males and females in size or shape are less than those found in all other domestic animals. The permanence of a pairing reduces the importance of pronounced sexually dimorphic traits that play a role in male competition for mating rights. Female dogs have as great a tendency to guard territory as males and vocalizations of males and females are similar. In wolves, males participate in feeding and caring for young as they are weaned from milk to almost the same degree as females. These are all characteristics that are important in promoting and maintaining an effective family unit when survival of offspring is as important to the male as to the female.

■

GENERAL ASPECTS OF SEXUAL BEHAVIOR

Precopulatory and Copulatory Responses

Before going into a discussion of the sexual behavioral characteristics of females and males and the particular responses of the individual species, we can make some observations about sexual behavior of domestic mammals in general. For example, it is useful when describing sexual behavior of both sexes to differentiate between precopulatory, copulatory, and postcopulatory responses. Also, we can talk about those stimuli (such as olfactory, visual, and auditory) that attract males and females to each other or that affect their interactions. Precopulatory behavior, often referred to as courtship, reflects the motivational aspects of sexual function. For the male this is pursuit of the female, licking and sniffing of the female genitalia, and mounting. For the female this may be seeking out males or otherwise demonstrating her impending estrous condition. In females this type of precopulatory behavior is referred to as proceptive behavior to differentiate it from receptive behavior, which refers to the more passive female reactions to a male's sexual advances (Beach, 1976).

Sometimes precopulatory stimulation of males can increase the sperm count of the ejaculate and enhance fertility. In females, precopulatory stimulation can also have an impact on reproductive performance. For example, the number of times a boar nudges the sides of a female before copulation can be directly related to litter size. Not only that, but when sows are artificially inseminated, they have larger litters if they are courted first by a boar (Hemsworth et al., 1977).

Copulatory responses are often referred to as consummatory responses, which culminate or terminate a sequence of sexual interactions. In males, the ejaculatory reaction is the primary consummatory response. Except for cats and dogs, it is difficult to identify any behavioral consummatory response in females. Female cats show brief periods of extreme arousal a few moments after intromission by the male. Similarly, the female dog shows a vigorous twisting and turning that also occurs a few seconds after the male's intromission.

Precopulatory and copulatory aspects of sexual behavior may also be differentiated on an ontogenetic basis. In males, mounting often occurs in the juvenile and prepubertal animals, whereas the ability to achieve erection and ejaculation does not generally occur before puberty. None of the behavioral aspects of female sexual behavior occur before puberty.

There is a great deal more variation among individuals of a species in precopulatory behavior than in consummatory responses. Some males may go through a much more elaborate or prolonged courtship than other males. Ejaculatory reactions vary only a little among individual males, and in fact the behavior is so stereotyped that ethologists tend to refer to ejaculation as a "fixed action" pattern.

Stimuli That Attract Males and Females to Each Other

Both males and females tend to respond to particular stimuli associated with the opposite sex. Although visual and auditory stimuli play a role in sexual behavior, most attention has focused on olfactory stimuli.

Olfactory Stimuli Vaginal secretions and urine of females in estrus contain sex attractants, or pheromones, which sexually arouse males and attract them to females. The chemical compounds constituting some of the sex pheromones of females have been isolated. In hamsters, dimethyl disulfide, found in the vaginal discharge of estrous females, has an attractive and sexually stimulating effect on males, but the potency of this compound is less than that of the vaginal discharge itself (O'Connell et al., 1978; Singer et al., 1976). The compound, methyl-*p*-hydroxy-benzoate, has been isolated from the urine of female dogs and is reported to have attracting and stimulating properties to male dogs. When the compound was applied to bitches that were sexually unreceptive, males attempted to mount them despite the female's resistance (Goodwin et al., 1979).

It is clear that the urine and vaginal secretions of estrous females smell different than the urine and vaginal secretions from females not in estrus. The human nose is apparently incapable of detecting these sex attractants, but they are apparent to animals even of different species. For example, rats can be trained to press a lever in response to air bubbled through urine from estrous cows and to refrain from pressing the lever when the air is from urine from nonestrous cows (Ladewig and Hart, 1981; Figure 3-1). This work is quantitative confirmation of old herdsmen's claims that some cow dogs can reliably pick out estrous cows from a herd without being instructed as to which cows to isolate (Akhlebininskii and Ishntov, cited by Jöchle, 1975). A laboratory demonstration of the ability of dogs to detect cows in estrus was recently conducted (Kiddy et al., 1978).

Males also produce sex pheromones, which attract and sexually stimulate females, but these have received less attention

Figure 3-1   Rats can be trained in an apparatus such as this to discriminate the smell of urine of estrous cows from that of diestrous cows. (From Ladewig and Hart, 1981).

than pheromones from females. A chemical scent has been isolated from the urine of the male red fox during the mating season (winter) that is not found in female urine. Since foxes are asocial, the reciprocal attraction of males to females is particularly important, and this male chemical may aid females in finding a potential mate (Jorgenson et al., 1978).

The best documentation of a male sex pheromone in domestic animals is in swine, in which it is observed that odors produced by boars have an arousal effect on sows and gilts that are in estrus (Signoret, 1970). Boars produce two sexual attractants, one in the secretions of the preputial pouch and another in the saliva secreted by the submaxillary salivary gland. During sexual excitement and precopulatory interactions with the sow, the boar produces copious amounts of foamy saliva that drips from his mouth. The active components of both secretions have been isolated; they are androgenic steroids with musklike odors (Aron, 1979). The chemical from the preputial gland is responsible for the taint in boar meat.

These attractants, or pheromones, which selectively activate certain neurons in the olfactory bulb (MacLeod et al., 1979), have been tested for their effect on the display of the immobility response of young female pigs (gilts) when a herdsman places pressure on their backs. Only about half of gilts in estrus pass this back pressure test in the absence of a boar, but when an aerosol containing the sex attractant from boars is sprayed toward the snouts of gilts in estrus, about 80 percent respond to the back pressure test (Melrose et al., 1971; Signoret, 1971).

Another example of the effect of male sex odors on females is found in cats. The presence of a male cat often evokes precopulatory responses in a female cat such as rolling as well as elements of the sexual stance, particularly crouching and treading of the back legs. Simply placing an estrous female in a cage recently occupied by a male will evoke the same responses (Michael, 1963).

Despite the importance of olfactory stimuli in sexually arousing both males and females, eliminating the ability of animals to smell does not eliminate their sexual responsiveness when confronted with a suitable sexual partner. This has been tested in male cats, dogs, sheep, goats, and swine (Aronson and Cooper, 1974; Hart and Haugen, 1972; Booth and Baldwin, 1980). Olfactory stimuli are obviously important in long-distance communication of sexual status and mutual attraction of males and females. Up close, however, visual and auditory stimuli convey sexual readiness and compensate for the lack of olfactory stimuli.

One could speculate that the sex pheromones of urine or vaginal secretions must be chemically quite different for different species. Males are sexually excited only by vaginal discharges or urine from estrous females of their own species, even though they may be able to detect the sex pheromones of a different species. If males were attracted to the sex pheromones of many species, they would spend all their time investigating one urine spot after another, or making sexual responses to females of a variety of species.

For a particular species some of the pheromones are volatile and probably easily detected by regular olfaction. Other pheromones appear to be nonvolatile and are normally detected by the vomeronasal system when a male nuzzles or contacts the genital area of a female or freshly voided urine. The vomeronasal organ is connected to the accessory olfactory system (Figure 3-2).

Flehmen Behavior   A response commonly made by male cattle, sheep, goats, horses, and cats to vaginal secretions and urine is

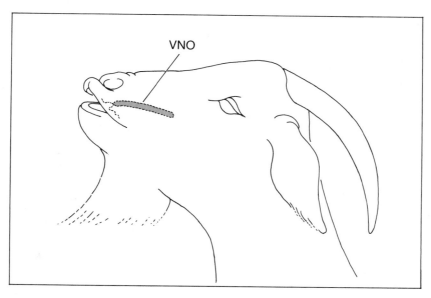

Figure 3-2   The vomeronasal organ (VNO) is the peripheral organ of the accessory olfactory system. During flehmen, fluid-borne nonvolatile chemicals, such as sex pheromones, can enter the vomeronasal organ through the oral cavity by means of the nasopalatine (incisive) duct.

a behavior characterized by head elevation and a curling of the upper lip, referred to as flehmen (Figure 3-3). Flehmen appears to be related to the functioning of the vomeronasal organ and the chemosensory investigation of nonvolatile sex pheromones of urine or vaginal secretions that are taken into the oral cavity. Experiments in goats reveal that during flehmen, fluid is drawn from the oral cavity through the nasopalatine duct and into the posterior part of vomeronasal organ, where materials can be subjected to chemosensory analysis (Ladewig and Hart, 1980) (Figure 3-4).

Flehmen behavior is evoked by the smell of urine or of the genital area; goats rendered anosmic by olfactory bulbectomy do not perform flehmen. Flehmen and the use of the vomeronasal system appear to be involved in confirming or refining olfactory discrimination by the regular olfactory system. The behavior is as likely to be performed whether the material is from estrous or nonestrous females. One important function of flehmen behavior is possibly to distinguish a female just coming into estrus from one not in estrus, a type of subtle discrimination that may not be possible with the main olfactory system (Hart,

Figure 3-3 Flehmen behavior in males is usually evoked by the male's close examination of the genital area or recently voided urine of females as seen here in goats, horses, cattle, sheep, and cats.

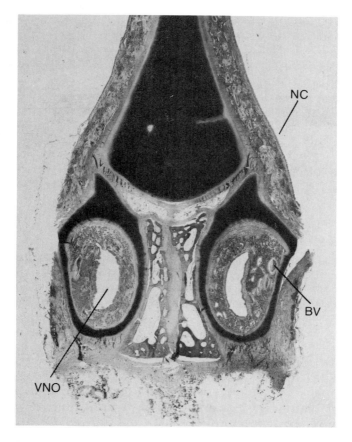

Figure 3-4 Low-power magnification of the nasal septum illustrating the vomeronasal organ. Abbreviations: VNO, lumen of the vomeronasal organ; BV, large blood vessel; NC, nasal cavity. (From Ladewig and Hart, 1980.)

1983). This concept is supported by the fact that when urine is presented to male goats, they perform close investigation and flehmen much more often in response to urine from nonestrous than estrous females (Ladewig et al., 1980) (Figure 3-5).

Flehmen is also performed by males in response to urine from other males or to the male's own urine. Thus the chemosensory information obtained by flehmen is functional in certain nonsexual as well as sexual situations. Even novel odors from inorganic compounds that an animal encounters in the pasture will occasionally evoke flehmen.

Flehmen is rarely performed by females during sexual encounters with males. Cows occasionally perform the behavior

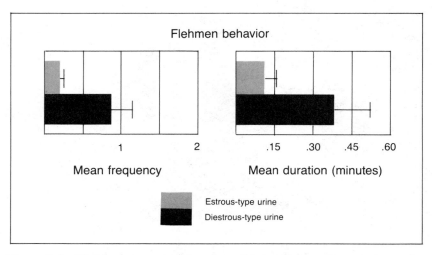

Figure 3-5 The frequency and duration of flehmen in male goats is much greater in response to urine from female goats in diestrus than estrus. One thought is that the male performs flehmen to check out whether the urine is from females in diestrus or proestrus. (From Ladewig et al., 1980.)

when sniffing other cows that are in estrus or proestrus (Esslemont et al., 1980; Hurnick et al., 1975). It is also performed occasionally by females, especially toward the birth fluids of newborn animals. The vomeronasal organ is quite extensive in nonmammalian vertebrates and is used in food detection, pursuit of prey, and sexual encounters. All mammals have a vomeronasal organ, with the exception of Old World primates, humans, and cetaceans. Some mammals such as dogs and rodents do not perform flehmen. These animals probably move materials into the vomeronasal organ by nuzzling, jaw movements, or head bobbing. In female rodents the vomeronasal organ is stimulated by volatile chemicals from males and has a role in influencing the neuroendocrine system's regulating reproductive cycles and the expression of estrus (Wysocki, 1979).

Auditory, Visual, and Tactile Stimuli The most noteworthy examples of auditory stimuli evoking sexual interest are caterwauling by tomcats and the muted meow uttered by a female cat when she is in the proximity of a tom. Sometimes this occurs when a female is locked inside a house but knows of the presence of a tom outside.

As in the case of olfactory stimuli, there are examples from the animal world of auditory stimuli that humans do not perceive. Male and female rats exchange ultrasonic vocalizations, some of which appear to sexually excite their partners and communicate sexual readiness and some of which indicate sexual refractoriness (Barfield and Geyer, 1972; Geyer et al., 1978a, 1978b).

Visual displays, especially the sexual posturing of mares and female dogs and cats, are proceptive responses indicating sexual interest on the part of the females. Perhaps the most notorious visual signal of sexual interest in humans is the eye contact engaged in by both women and men in the early stages of social encounters.

Tactile stimuli from males are important in evoking sexual signals from females. Many small animals, such as rats, rabbits, guinea pigs, and hamsters, display a receptive posture in response to mounting by the male. The neck grip of the male cat activates the receptive posture and movements of the hind legs of females. Biting on the neck and withers, received by mares when a stallion mounts, may be sexually stimulating to the mares. Rubbing the flanks and genitalia of mares, whether done by the stallion or a human handler, may evoke behavioral signs of estrus from a mare that otherwise shows no signs of estrus. Chin resting of a bull on the back of a cow just prior to mounting may have some stimulatory effects on the cow.

Usually animals experience the influence of stimuli of more than one type, such as when a male watches a conspecific mate with a female. It has been documented in both beef cattle and dairy goats that watching another male mate enhances sexual performance in the spectator when he is allowed to mate (Mader and Price, 1984; Price et al., 1984). Interestingly, watching another male mate was more stimulatory than being watched by another male or just being restrained near the female.

SEXUAL BEHAVIOR OF FEMALES

The most prominent aspect of female sexual behavior is its cyclic nature. With few exceptions (among the primate species), females display little sexual interest in males and will not be receptive toward a male's sexual advances outside the estrus pe-

Figure 3-6 Lack of receptivity is often displayed by a mild aggressive response of the female directed toward the male, as seen here in sheep.

riod. During estrus the olfactory, visual, and sometimes auditory stimuli produced by females have obvious attractive qualities for males. In the absence of some species-typical sign of estrus, especially the presence of a sex pheromone, males of wild species will not make sexual advances toward females. This is also evident among most dogs, cats, and livestock species under free-ranging conditions. However, males that are frequently used for stud service will commonly attempt to mount females that are not in heat. One finds with stud males a tendency for them to "test" females by attempting to mount on first exposure to them. If the female is not in estrus, she usually will not "stand" and often repels the male by aggressive responses (Figure 3-6).

The demands of efficient reproductive performance in farms have reduced our chances of observing the behavior of females seeking out or choosing certain males. When given the opportunity, ewes will seek out rams when in estrus (Lindsay and Fletcher, 1972). The tendency for female dogs and cats to seek out sexual partners when in estrus is also well known by pet owners. The preference of female dogs for particular males, to the extent where some bitches are perfectly receptive to some males but utterly refuse to mate with others, has been well documented experimentally (Beach and LeBoeuf, 1967).

Estrous Cycles

Domestic species have a variety of types of estrous cycles. As mentioned before, characteristics of the estrous cycles relate to the reproductive pattern of the wild ancestors. Polyestrous

TABLE 3-1 Estrous cycles of domestic animals

Species	Type of estrous cycle	Cycle length	Duration of estrus
Dog	Monestrous	One cycle each season (two seasons per year)	7–10 days
Cat	Seasonally polyestrous	3 weeks	1–4 days with mating; 7–10 days without mating
Horse	Seasonally polyestrous	3 weeks	5–6 days
Cattle	Polyestrous	3 weeks	10–12 hours
Sheep	Seasonally polyestrous	16–18 days	36 hours
Goat	Seasonally polyestrous	3 weeks	12–24 hours
Swine	Polyestrous	3 weeks	2–3 days

species cycle several times during the breeding season; thus, if fertilization does not occur during one estrus period, there is another chance within a few weeks again.

The estrous cycles of domestic species are categorized in Table 3-1. Cattle, swine, sheep, goats, horses, and cats are polyestrous, with the duration between estrus periods varying between 10 days and 25 days. Sheep, goats, and cats are the most seasonal in their display of polyestrous cycles. Horses tend to be seasonal, but some mares cycle for a long time much of the year. In cattle and swine seasonality seems to have been virtually eliminated by artificial selection. Goats and sheep are interesting in that in temperate zones they are seasonally polyestrous; but in the tropics, where environmental conditions allow survival of young year-round, they cycle throughout the year.

Although dogs cycle twice per year at roughly six-month intervals, they are referred to as monestrous because their wild ancestors and relatives cycle only once a year. Some wild canids, including the wolf, are monogamous, and with the male continuously around, there is less chance of a missed estrus period. The duration of estrus is longer in dogs than other species, allowing for more copulations during the single estrus period.

Behavioral Signs of Estrus

Females show a variety of species-typical behavioral signs of estrus or impending estrus. The signs vary according to whether females are normally around males or isolated from them and whether or not they are with other females. Dairy cows, for example, are usually artificially inseminated and are therefore rarely exposed to bulls. There are behavioral signs shown around other cows that give some indication of estrus. Mares kept away from stallions often display no signs of estrus until exposed to a stallion.

As we learn more about the exact types of stimuli from males that evoke signs of sexual interest in females, it is becoming possible to duplicate these artificially as an aid in the detection of estrus. The effects of spraying a sex pheromone at a gilt together with back pressure in evoking signs of estrus is a classic example. Another example is the success Veeckman and Ödberg (1978) have had in evoking signs indicative of estrus in mares (lifting of the tail, spreading of the legs) by playing recordings from a courting stallion and rubbing the flank and external genitalia (stimulating the stallion's nudging).

The practical significance of these procedures is obvious when one must rely on artificial insemination. However, it is also of value in determining when to bring the female to a male for breeding.

Changes in Sensory and Motor Behavior   Females of almost all species appear to show a marked increase in general activity when they come into estrus. For cattle, sheep, and horses, this may mean an increase in milling around, exploration, vocal activity, or agonistic behavior toward other females. In laboratory rats a marked increase in wheel-running activity (in caged animals) is a reliable quantitative indicator of estrus. No female seems to escape this energizing effect of estrogen. In one study on women in which general movement about the house was monitored by pedometers, there was an increase in walking associated with the time of ovulation (Morris and Udry, 1970).

One of the more interesting changes in human females at the time of ovulation is an increase in general sensory acuity during the time when estrogen levels are highest (Diamond et al., 1972). This has been documented for both olfactory and visual acuity. The degree to which such changes may occur in domestic animals has not been explored. It has been argued that increases in sensory acuity and awareness relate to an increased tendency toward copulatory activity.

Behavior of Estrous Females Toward Other Females

When females are maintained in a herd, an increased level of agonistic interactions with conspecifics is often observed when one or more females come into estrus. This is most notable in mares.

In cattle especially, but also to a limited degree in pigs and dogs, there is a tendency for females to mount those that are in estrus (Beach, 1968).

The detection of estrus is particularly important in dairy cattle, in which artificial insemination is routinely used. As a cow approaches estrus, she becomes more restless and walks around more even at the cost of eating less. She engages in more butting of other cows, approximately doubling the normal frequency. She attempts to mount other cows at times but until she is in estrus does not stand when she is mounted by other cows. Observations on individual cows suggest that when a cow approaches the stage of proestrus, she tends to search for a partner by nudging resting cows. If no cohort can be found, she may mount a cow in diestrus. If another cow in proestrus is found, they may engage in mutual sniffing the genitalia, playful head-to-head fighting, and mounting (Hurnik et al., 1975). The most reliable sign of the 10-hour period of estrus is that of standing while being mounted by other cows (Esslemont et al., 1980; Hurnik et al., 1975). The cows doing the mounting are also likely to be those in estrus or approaching it (Mylrea and Beilharz, 1964). Before mounting another cow, a cow may sniff the genital region and sometimes perform flehmen.

Mounting among cattle is the single most frequently used indication of estrus. Cows in estrus tend to associate in groups with other estrous cows, especially at night, and mounting activity in these groups is particularly noticeable. Observations of cows in the field have been found to be more effective in detecting behavioral signs of estrus than observations in the milking parlor (Williamson et al., 1972). As estrous heifers concentrated themselves into a sexually active group and roamed extensively about the pasture together, proestrous heifers were attracted into the group. To detect estrous heifers, bulls had only to locate the sexually active group, which they did visually rather than by olfactory cues (Blockey, 1978).

A study on mounting activity in dairy cows found that cows were mounted an average of 56 times during an estrus period of 10 to 15 hours. However, there was considerable variability in the frequency with which they were mounted (Esslemont and Bryant, 1976). Like Williamson et al., Esslemont and Bryant documented that mounting was most likely to occur in feeding and rest areas and least likely to occur in the holding area near the milking parlor, which is traditionally the area where farmers attempt to observe mounting.

Some commercially available devices can be applied to the rump or tailhead region of cattle that when broken by the pressure of other cows mounting change color. Rubbing a crayon over the rump of cows expected to come into heat and periodically checking for smearing of the crayon mark is another procedure used to detect estrus (Foote, 1975).

Female sheep and horses show few behavioral signs associated with estrus or impending estrus in the absence of males. If adult sexually active males are in the vicinity of ewes, the behavioral signs of estrus are often obvious. When one wants to employ artificial insemination for sheep, or when the herd manager wants to know what females have been bred, a marking device may be attached to the front of a ram so that he marks each ewe he mounts. Also, some sheep owners check ewes daily with vasectomized rams.

As mentioned above, most dairy cows do show changes related to estrus when apart from males and in the presence of other cows. This behavior is used in the timing of artificial insemination. However, a small percentage of cows do not display these behavioral signs very prominently, and undetected estrus is an important economic problem in dairy cattle reproduction. Cows that give little behavioral indication of estrus around other cows are said to have a silent estrus. Careful analyses have indicated that mounting and increased activity of estrous females may easily be missed unless the herd manager spends an hour or so in the late evening, after the cows have settled down, carefully observing the animals. It has also been noted that running with the cows a vasectomized bull or a steer injected with testosterone often increases the ability of an observer to pick out estrous animals.

The value of mechanical techniques to indicate that mounting has occurred is documented by a study of cows that, according to the herdsman, showed no behavioral indications of estrus. The application of heat mount detectors revealed that the majority of these cows were actually mounted by other cows, but at a time or place that was missed by the herdsman (Schels et al., 1978).

The marked tendency of cows to display mounting behavior as an aspect of estrus could well be a reflection of artificial selection. If a cow emits strong olfactory and visual signals and approaches other cows making it easy for them to mount her, she is likely to be mounted. Those cows that attract the most mounting, and that also engage in mounting themselves when in estrus, are the ones most certain to be artificially inseminated by a herdsman who relies on such behavioral signs for breeding cows. The most sexually demonstrative cows produce the most offspring.

The estrous behavior of ewes and mares, in which they display only to conspecific males but not to other females, is closer to the natural behavior of wild ancestors. Cows that go through silent estrus—that are not mounted by other cows and do no mounting themselves—are the bane of dairy breeders and a primary cause of financial loss from missed estrus periods, but they are probably the most normal cows from the standpoint of natural behavior.

Behavior of Estrous Females Toward Males   Proceptive responses are quite pronounced in females of some species when they are given free rein to search out males. In most instances we keep females confined, so that their options are limited. With sheep it is still customary, however, for ewes to be run in a pasture with one or more breeding males. Under these circumstances, ram-seeking behavior is quite pronounced. A ram can even get solicitations from two ewes simultaneously who fight over him in an attempt to preempt his sexual favors (Hulet et al., 1962b).

Apart from proceptive behavior, females show a variety of behavioral patterns in the presence of males that indicate receptivity. One of these patterns is the tendency of females in estrus to urinate in the presence of a male (Figure 3-7). This behavior is prominent in horses, sheep, and goats, but rather infrequent in cats, small laboratory animals, and primates. Sometimes female dogs display an increased tendency to urinate in the presence of a male. Not infrequently bitches will use a

Figure 3-7 Urination in the presence of a male is a behavioral sign of estrus in the mare and the ewe, as shown above.

posture suggestive of the leg-lift commonly associated with urination in adult male dogs.

Since the urine of estrous females contains sex pheromones, the behavior of urinating in the presence of a male has the value of communicating to the male the female's state of estrus. The behavior is also a visual display that a male can detect before he is within olfactory range. Associated with the tendency of mares to urinate in a male's presence is their behavior of everting the clitoris rhythmically four or five times in the male's presence.

Horse breeders refer to this as "winking." In horses, winking, tail raising, squatting, and urinating, in the absence of kicking, are the best indications of estrus (Back et al., 1974).

Although olfactory, visual, and auditory stimuli are important distance communicators of sexual status, to human observers the most obvious indication of a female's receptivity around males is her willingness to accept a male's sexual advances and to stand when he attempts mounting. In fact, it appears as though that is about all some of the stud males go on. Signoret (1975) found that attempts by rams to copulate with restrained ewes were determined more by their degree of immobility than anything else. If a ewe in anestrus was restrained the ram would still attempt copulation. The same behavior by males toward restrained females has been observed in dairy cattle. Blockey (1981a, 1981b, 1981c) has developed a serving capacity (mating performance) test for beef bulls in which restrained females are used as mounting stimuli. Bulls must first be sexually stimulated by observing bulls mounting cows or cows mounting each other. Part of male nondiscriminatory behavior may reflect the fact that they are highly conditioned to mating in a certain location.

The standing behavior has received particular attention in pigs and is called the immobility response because it can be evoked by humans placing pressure on gilts' backs. The effect of boar odor in facilitating this behavior has been mentioned above.

If a male is slow in initiating sexual advances, or gets distracted, females of some species may mount the male. This is common, for example, in cattle and dogs (Figure 3-8). Often the mounting is accompanied by pelvic thrusting as intense as that shown by males during copulation.

Dogs engage in a behavior referred to as teasing, in which the female shows a willingness to stand and accept the male's sexual advances but interrupts this behavior by dashing off for a short distance, and then stands again.

The willingness to stand for a male's sexual advances is, for dogs, cats and rodents, related to the adoption of a special posture and movements. Such special postures are also characteristic of the receptive behavior of rats, hamsters, guinea pigs, rabbits, and primates. In the case of dogs and cats, these postures are known to be partially spinal reflexes that are controlled at the spinal level by estrogen (Hart, 1978a, 1978b). The postures

Figure 3-8 Mounting of males by females is sometimes displayed by cows in estrus.

are usually evoked by the tactile stimuli the female receives when mounted by the male (Figure 3-9). The receptive posture may be shown by females prior to mounting by the male.

In dogs the posture and movements are usually not easily evoked by human handlers, but in female cats, the pelvic ele-

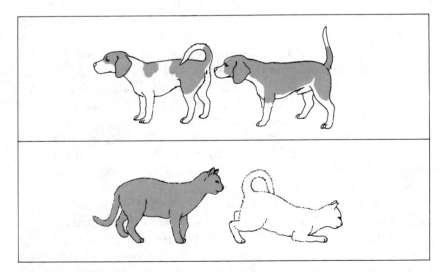

Figure 3-9 The receptive posture of dogs and cats is displayed by females upon the approach of males. The posture and movements characteristic of female cats in heat can also be evoked by human handlers.

vation, treading of the back legs, and lateral deviation of the tail, which are characteristically shown in the presence of toms, can often be elicited by a person stroking the back and/or gently touching the perineal region.

A lack of willingness to stand or to show other behavioral signs of receptivity indicates that the female is not in estrus. At times, however, a female may be physiologically in estrus (as can be determined by blood hormonal analyses or vaginal smears) but display no behavioral signs of estrus, even to the point of not allowing a male to mount. This may reflect a particular dislike for certain males or fear of males. At other times the phenomenon is simply a manifestation of hormone-sensitive areas of the brain not responding to estrogen secretions the same as the rest of the body.

Female Responses during Copulation

Do female animals have orgasms? It is difficult to find examples of behavioral and physiological responses in animals that are possibly analogous to orgasmic reactions in human females and males. Clearly some internal physiological responses occur during copulation. For example, in cows increased uterine tone and extensive contractions have been noticed within seconds following copulation. Interestingly, such uterine motility also follows nuzzling and mounting by a bull. The uterine motility, which probably plays a role in rapid transport of sperm, is attributed to the release of oxytocin (Van Demark and Hays, 1952).

The best examples in female animals of outward manifestations of reactions comparable to orgasm are seen in cats and dogs. Five to 10 seconds after intromission by the male cat, the female cat typically becomes very aroused, her eyes dilate, and she usually turns and hits at the male, causing him to promptly dismount and move away. Immediately thereafter she engages in a bout of rolling and rubbing on the floor or ground for several minutes. The rolling and rubbing, referred to as the copulatory afterreaction, looks as though it is pleasurable. However, if during this time the male approaches the female, she is likely to hit at him.

The copulatory behavior of dogs differs from other species in that there is a genital lock or tie of from ten to twenty minutes. The female dog exhibits a copulatory reaction that looks like an orgasmic reaction, but in a different way than that of the cat.

About 10 seconds after intromission begins the bitch goes through an excited, rigorous twisting and turning reaction. The reaction begins when the bulb of the penis reaches maximum size and presumably stimulates the clitoris. Rhythmic contractions of the vaginal constrictor muscle also occur during the twisting and turning, due to stimulation of the clitoris (Hart, 1970).

One characteristic of sexual activity that has not received much attention in females is the postcopulatory refractory period. Females will usually engage in sexual activity in a shorter period of time after copulation than males, so that a refractory period is not evident unless one mates a succession of males with females.

■

SEXUAL BEHAVIOR OF MALES

In the wild ancestors of domestic species, breeding strategies of males centered around who bred which females and the most females at the right time. Hence, courtship behavior, male-male competition for females, and the female's attempt to attract the strongest males were very important. On the domestic scene, we are less concerned with these lofty standards and have artificially selected males and females to do our bidding. Hence, courtship and precopulatory responses represent interference with efficient farm breeding. It is common to see courtship behavior considerably abbreviated in comparison with that of wild ancestors. The act of copulation, though, probably remains fairly true to form. In wild ruminants copulation takes only a second or two. This very short duration probably evolved to reduce the risk of predation during sexual encounters. In what follows I shall discuss the male sexual behavior that is typically displayed under domestic conditions. Figures 3-10 through 3-16 portray the distinctive characteristics of male and femal sexual behavior for the various domestic species.

Precopulatory Responses

Males of almost any species, when placed with a sexually receptive female, usually first investigate the genitalia of the female. This may involve licking of the genital region and smelling

Figure 3-10 Copulatory behavior of dogs: A, The female exhibits lateral curvature and tail deviation as the male investigates her anogenital region. B, The male next engages in mounting and pelvic thrusting. C, Intromission is followed by 15 to 30 seconds of intense pelvic movement and leg stepping (intense ejaculatory reaction), which precedes the genital lock or tie. D, Immediately after the onset of the tie, the female displays vigorous twisting and turning and may even throw the male. Experienced males usually dismount before they are thrown. E, After the dismounting the female may continue twisting and turning. F, Both the male and female remain quiescent during the lock, which usually lasts from 10 to 30 minutes.

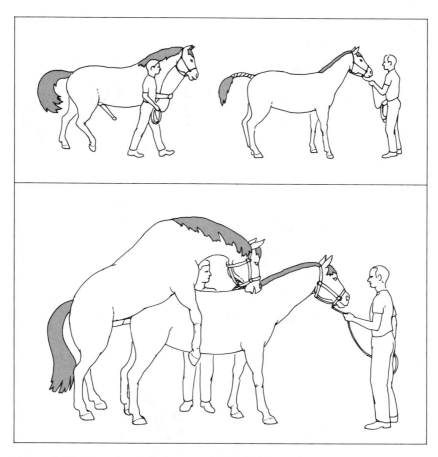

Figure 3-12 Copulatory behavior of horses: The stallion approaches showing considerable sexual excitement and an almost completely erect penis. The mare may urinate in the stallion's presence at this time. Mounting is followed by trial-and-error thrusting until intromission is achieved. Often the stallion will grasp the neck or withers of the female with his teeth. Ejaculation takes about 5 to 15 seconds. Some vertical, rhythmic tail movement and leg stepping may occur during ejaculation.

Figure 3-11 Copulatory behavior of cats: A, The male approaches the female while she exhibits pelvic elevation, treading of the back legs, and tail deviation. B, The male takes a neck grip, mounts the female, and engages in some leg stepping as he slides backward on her. C, Thrusting leads to intromission and a deep pelvic thrust by the male lasting from 1 to 3 seconds. D, The female becomes overtly aroused, emits a copulatory cry, and usually turns on the male, scratching at him. E, The female next licks her genital region. F, Several minutes of rolling and rubbing on the floor characterize the postcopulatory afterreaction of the female.

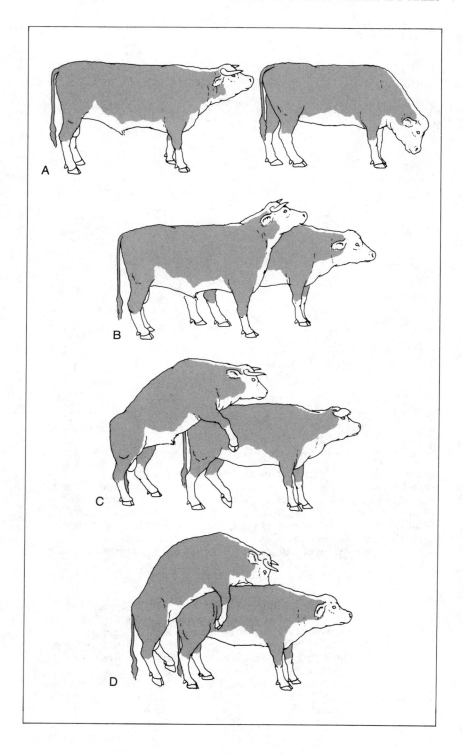

of voided urine if the female urinates in the male's presence. Flehmen is often performed by cattle, sheep, goats, and horses on investigation of the anogenital area of females or voided urine. Flehmen is performed by male pigs and cats, but it is less prominent.

In dogs there may be considerable time devoted to playful interaction with the female, particularly if the female acts in a playful or teasing manner herself. The next response of the male is usually to mount and attempt copulatory intromission by pelvic thrusting. In horses, dogs, and cats intromission occurs as a result of trial-and-error thrusting. In livestock species intromission generally follows a well-directed thrust.

There are individual differences in regard to the intensity and duration of precopulatory sexual behavior. In some males this aspect of sexual activity is prolonged whereas in other males attempts at copulation may occur with practically no investigation of the female. There are also marked differences between some breeds, especially in cattle. Beef bulls tend to engage in much more precopulatory investigation and courtship activity than dairy bulls.

The prolonged courtship of beef cattle is undoubtedly more natural and remains that way because most breeding is allowed to occur under pasture conditions. Breeding of dairy cattle has been a different proposition because cows are usually kept under closer confinement. If natural breeding is done at all, the farmer does not want to waste time waiting for a bull to go through the preliminaries. Besides, dairy bulls are dangerous and difficult to handle. Hence we have selected dairy bulls, over the years, from sires that tended to omit foreplay. Of course, with artificial collection of semen there may be no cow for miles around. We use bulls that will rapidly mount anything that stands, including other males.

Males also differ in the degree of adaptation or familiarity with the environment that they need before they will initiate

---

Figure 3-13 Copulatory behavior of cattle: A, Many sexual encounters begin with genital investigation, which is often immediately followed by flehmen by the bull. B, A mount of the female is frequently preceded by the male placing his chin on the female's back (referred to as chin resting). C, Mounting by bulls is followed immediately by ejaculation, trial-and-error thrusting usually does not occur as in horses. D, Ejaculation occurs during one powerful pelvic thrust in which the bull's back legs are sometimes lifted off the ground by muscular abdominal contractions. Ejaculation lasts only a second or two.

Figure 3-14 Copulatory behavior of sheep: A, A sexual encounter often begins with genital investigation, after which the ram may perform flehmen. Ewes that are in estrus often urinate on the approach of the ram. B, A ram frequently nudges the ewe several times before attempting to mount. In nudging, the ram brings his shoulder into contact with the ewe's flank and makes short, chopping kicking movements with a foreleg. The ram's head is tilted sideways, and he usually utters a low-pitched vocalization during nudging. C, Mounting is very brief if not followed by intromission. D, During a mount when intromission and ejaculation occur there is a backward head bounce. Ejaculation lasts only a second or two.

Figure 3-15 Copulatory behavior of goats: A, After genital investigation the buck may nudge the female about the head and neck. B, Another type of nudging involves the male directing short, choppy kicking movements of the foreleg toward the female. The buck may also lower and tilt the head sideways toward the female while uttering low-pitched babbling vocalizations. C, Mounts without intromission are brief. D, Like rams, bucks display a backward head bounce and arching of the neck during ejaculation. Intromission and ejaculation are characterized by a deep penile thrust and backward retraction of the forelegs. As with cattle and sheep, ejaculation lasts only a second or two.

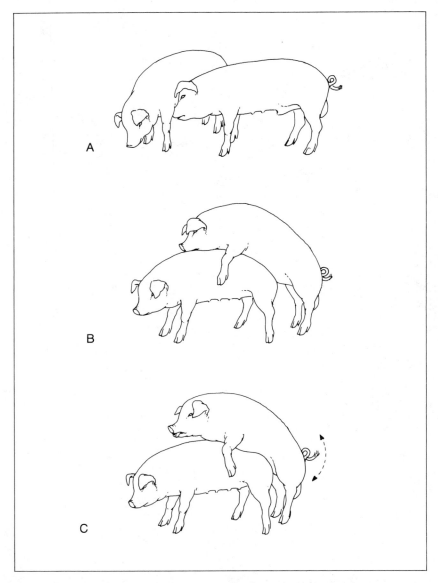

Figure 3-16 Copulatory behavior of swine: A, Nuzzling by the boar of the sow is a prominent precopulatory behavior. The nuzzling is done with the snout and is directed toward the head, shoulder, side, and genital region of the sow. During this precopulatory phase, the boar grunts almost continuously, grinds his teeth, and foams at the mouth. B, Mounting is followed by trial-and-error shallow pelvic thrusts and extensions of the penis. C, Ejaculation, which commences after intromission is achieved, is very prolonged, ranging from two to 30 minutes. Little thrusting occurs during ejaculation; the main outward sign of ejaculation is rhythmic vertical movements of the tail.

sexual advances. This is recognized in all species but is well documented with regard to laboratory cats. Some toms require almost a month of repeated exposure to a new environment before they become consistent breeders. This tendency is not as pronounced in dogs and large animals, but adaptation of at least a few days may be required. For some reason, strange environments appear to have less inhibitory effects on females, or at least they can be bred while exploring. Breeding is usually most successful if the female is brought to the male's familiar home area rather than the converse.

Copulatory Responses

Erection Erection is obviously necessary for copulation and the deposition of sperm into the vagina of the female genital tract. In some instances there is evidence that erection facilitates the introduction of sperm in the uterus. In ruminants extrusion of the penis occurs simultaneously with a strong pelvic thrust and a peristaltic contraction of the smooth muscle lining the urethra. The result is that sperm may enter the uterus almost by pure force. There are other more gentle, but just as effective, mechanisms of ensuring that sperm reaches the uterus. In pigs, the penis has a corkscrewlike tip that is threaded into the cervix; the cervix has grooves that correspond to the corkscrew, and thus seminal material flows directly into the uterus. In cats the penis is covered with cornified spines, which are directed backward. A function of the spines might be to keep the penis from being forced backward until ejaculation has occurred.

In dogs, the fibroelastic erectile tissue of the root of the penis becomes engorged prior to copulation, resulting in a stiffening of the penis from the pelvic bone to the cartilaginous tip of penile bone. This allows intromission to occur. However, following intromission there is engorgement of the expansive erectile tissue. The bulbous portion of the penis is so engorged that it cannot be pulled from the vaginal orifice. Hence, the animals are tied until spinal reflexive mechanisms that maintain this engorgement "run down." Since the penis completely fills the vaginal lumen, when ejaculation occurs, all the seminal material is forced into the uterus because there is no room in the vagina (Grandage, 1972).

The one animal in which erection occurs long before mounting is attempted is the stallion. The anatomy of the penis in the stallion is quite similar to that of the human male. Perhaps this similarity in anatomy and sexual arousal is why authors of ancient myths tended to use the stallion as a symbol for human male sexuality. The anatomical similarity is rather superficial, however. Just after intromission, and just before the onset of ejaculation, the rim of the coronal part of the tip of the penis expands into a cuplike shape that seals off the vagina and presumably causes seminal material to be deposited in the uterus.

Ejaculation   Intromission is almost always followed by a species-typical ejaculatory response closely resembling that of the wild ancestors. This response is believed to be a spinal reflex evoked by pressure applied to the body of the penis. This, of course, usually occurs just after intromission, but it may occur extra-vaginally if the male's thrusting attempts apply pressure to the penis. Domestic animals differ from man and other primates in which ejaculation may not occur for several minutes after copulation begins.

Ejaculation is caused by the contraction of smooth and striated muscle along the genital tract. Depending on species-typical ejaculatory patterns, the contractions may range from one powerful peristaltic contraction, as in cattle and sheep, to a series of contractions lasting several seconds, as in horses and dogs.

It is perhaps instructive to categorize ejaculatory patterns of domestic animals into three types. In ruminants ejaculation lasts only a second or two. The behavior consists of one deep pelvic thrust accompanied by intense contraction of the musculature of the entire body. The response is of the "all-or-nothing" type. Sheep and goats display a backward head bounce as part of the ejaculatory reflex (Banks, 1964).

A second type of ejaculatory response is evident in horses. Pelvic thrusting during intromission usually lasts as long as 5 to 15 seconds. The response is all or nothing in the sense that there is a definite onset and termination of the response, but the duration may vary somewhat from time to time. Rhythmic contractions of one or more of the penile muscles continues for the duration of the response. Such contractions are often evident from the visible contractions of the anal sphincter, which contracts simultaneously with the penile muscles.

A third type of ejaculatory response occurs in pigs and dogs. In pigs, weak pelvic thrusting occurs over a period of from five to 30 minutes. The ejaculation is not all or nothing in that there is no definite termination. In one mating a boar may ejaculate for three minutes and in another for ten minutes. Occasionally the boar may ejaculate for a few minutes, dismount, and walk around a little, then intromit again and continue ejaculating. During ejaculatory contractions in swine there are contractions of the anal sphincter muscle and vertical movements of the tail.

Copulation in dogs deserves special comment because of the genital lock. In dogs and wild canids there is initially an intense ejaculatory reaction of from 15 to 30 seconds followed by a genital lock in which ejaculation continues for usually 10 to 30 minutes, but sometimes for as long as 60 minutes. Seminal fluid expelled during the lock contains mostly prostatic material, and the lock is not necessary for a fertile mating although it appears to enhance fertility. The lock is due to the maintenance of erection by a spinal reflex in the male (Hart 1978a, 1978b). The adaptive advantage of the lock for wild canids is not entirely clear. If canids were easily preyed on, the lock would be disadvantageous by making them more susceptible to predators. Because canids are not prey, they often have lots of time on their hands (paws), and being tied up for a while during the mating season is of no primary disadvantage. One possible selective advantage is that since sperm are probably deposited directly in the uterus, the lock allows a particular male's sperm prolonged time for access to the ova. Because of the headstart of the sperm from the first male, sperm from a subsequent male are at a competitive disadvantage.

Postcopulatory Refractory Period

Following ejaculation there is a period of sexual quiescence in which the male appears to have no sexual interest in females and is incapable of sexual activity. This period of sexual quiescence is referred to as the postcopulatory or postejaculatory refractory period. The refractory period consists of physiological and behavioral components. The physiological component is relatively invariant for each species and is related to refractoriness of spinal and brain mechanisms mediating the ejaculatory reflex (Hart, 1978b).

The behavioral component of the refractory period is quite variable for each male and is related to sexual excitement in the male. The best example of the degree of variability in the behavioral refractory period has been demonstrated by some work on cattle and sheep. When a bull or ram is left with a receptive female for a period of time, he will have a number of ejaculations, say five or six, and then lose interest in the female. After each ejaculation the interval between the occurrence of the next bout of sexual activity becomes progressively longer. Eventually the point is reached where the animal seems to be sexually exhausted. If a new female in estrus is introduced at this point, the male will usually show renewed sexual interest and ejaculate a number of times again (Bermant et al., 1969; Beamer et al., 1969; Pepelko and Clegg, 1965; Thiery and Signoret, 1978). The behavioral component of the refractory period is changed but the physiological component remains about constant. Changing the environment only by placing the same female in a new stall may also revive the male's interest (Almquist and Hale, 1956).

The sexually reviving effects of a new female have been shown for males of several species, most recently in rhesus monkeys. Males that were paired with the same female for about four years showed a decline in sexual activity. This deterioration in potency was abruptly and completely reversed by substituting a new female (Michael and Zumpe, 1978). The adaptive value of the novel-female effect in wild species is that it causes males to attempt to inseminate a greater variety of females than they might otherwise. The work on monkeys might have implications for domesticated human society with institutionalized monogamy. Michael and Zumpe wonder if this phenomenon can explain the notorious tendency of men to break and remake consort bonds with new partners. Might the phenomenon also be the basis of cultural means for periodically changing one's stimulus properties by the use of clothing, adornment, coiffure, and odor?

There are pronounced species differences in the typical duration of the refractory period and in the capacity for multiple ejaculations in a short period of time. Cattle, sheep, and goats are usually capable of ejaculating three to five times in one hour if the male is left with the same female for an extended period. In one study it was observed that rams ejaculated an average of 5.5 times in a test session with one female (Pepelko and Clegg, 1965). Over a period of several days this activity would slow down. A mean of 45 ejaculations in six days was recorded in rams by Hulet et al. (1962a), with the most active ram in their group mating 66 times in six days.

After the ruminants, cats have the most vigorous capacity for multiple ejaculations. Three ejaculations in an hour is not uncommon for tomcats. Horses, dogs, and swine seem to be capable of about one ejaculation per hour and no more than two ejaculations per day, although there are occasional reports of as many as four or five matings in a day.

In those species in which males have the capacity to ejaculate quite frequently, it is known the semen quality, as gauged by the number of sperm per ejaculate, decreases after several ejaculations. For example, it has been determined that the number of sperm per ejaculate in rams after the eighth ejaculation can fall below that required for fertilization (Jennings and McWeeney, 1976).

It appears that in species that characteristically ejaculate less often sperm can be depleted more easily. It has been estimated that using a dog for breeding once a day can deplete sperm reserves. An optimal concentration of sperm was obtained by using a dog at stud every two days (Boucher et al., 1957).

Artificial Collection of Semen

When attempts are made to collect semen and the animal is not accustomed to the manual collection procedure, or is in a strange environment, there is a good possibility that the ejaculatory response will be so strongly inhibited that it cannot be evoked by manual stimulation of the penis or even with the use of an artificial vagina. Placing a sexually receptive female near the male or allowing the male to mount a female a few times may result in sufficient sexual excitement that the ejaculatory reflex is released from this inhibition and semen collection attempts are successful. Where the inhibition may be partial, it has been noted that the quantity and quality (sperm content) of semen obtained appears to reflect the degree to which the ejaculatory reflex is released from inhibition. In cattle and dogs it has been documented that teasing a male prior to artificial collection of semen will increase the sperm count of the ejaculate (Almquist, 1973; Boucher et al., 1957). This is apparently not true in stallions (Pickett and Vos, 1975).

■

DISPLAY OF HOMOTYPICAL AND HETEROTYPICAL SEXUAL BEHAVIOR

At times when we attempt to generalize from our concepts about human sexual behavior to behavior of domestic animals, the picture of what is normal or abnormal can get distorted. This is particularly true in considering behavioral patterns typical of one sex (homotypical) that are displayed by the opposite sex. Both sexes have the capacity and the neural circuitry to display species-typical behavioral patterns characteristic of the opposite sex (heterotypical). Sometimes a display of mounting by females or a feminine receptive posture by males is part of the normal repertoire of wild animals. In domestic animals this may not be a reflection of the normal repertoire of the wild ancestors but may stem from the effects of artificial breeding or husbandry environment. We will discuss here the display of heterotypical sexual behavior when females engage in mounting of other females or males and the acceptance by males of being mounted by other males.

Mounting by Females

We have seen in the discussion of female sexual behavior that there is a pronounced tendency of cows to mount other cows in estrus. This is normal behavior, at least in the context of a farm environment, and the behavior is the best predictor of the state of estrus in the cow that is mounted.

More extreme forms of male behavior by females can be induced by unusual hormonal treatment. After prolonged treatment with androgens or estrogens, female sheep and rats have been observed even to display an ejaculatory response when mounting other females (Fabre, 1977; Emery and Sachs, 1975). A condition referred to as nymphomania, seen in cattle, horses, and other species, is sometimes characterized by females acting like males. Actually the cause of the syndrome is usually an ovarian cyst that produces estrogen for a prolonged period of time.

Sexual behavior typical of either sex is sometimes seen in a nonsexual context. When two female dogs are housed together, one may frequently express dominance over the other by mounting it. Such mounting usually differs from sexual mounting in that it is more to the side and is accompanied by growling. In a number of primate species both males and females may use the feminine presentation posture to indicate a subordinate position to a dominant animal. The dominant animal may even briefly mount the subordinate one.

Inappropriate Mounting by Males

Sexual responses may be part of play in juvenile or prepubertal individuals. This activity is particularly common among dogs. Mounting activity shown by juveniles may even consist of some pelvic thrusting. Female juveniles or prepubertal individuals only rarely mount other males or females.

Buller-Steer Syndrome   Although the secretion of testosterone at the time of puberty is needed to bring out the full display of male behavior, bull calves do have the basic neural circuitry for masculine behavior at birth. Castration does not alter the basic neural connections, and under certain situations, prepubertally castrated calves will display bull-like behavior. Mounting, usually in brief and random encounters, is commonly seen among groups of steers, especially those kept in close confinement as in feedlots. It has been reported that, given a choice, steers will almost always mount heifers rather than steers (Lesmeister and Ellington, 1977).

In most confined groups of steers, a few animals seem to be particularly attractive for others to mount. These steers are referred to as bullers. The attractiveness of the buller may be due to his solicitous behavior toward other males or possibly to the secretion of a noticeable amount of estrogen, which would act as a sex attractant. Also contributing is the fact that the buller does not butt away steers that attempt to mount him. Mounting by one steer stimulates others to mount or ride the buller, so that sometimes one may see a group of riders waiting their turns to mount.

The buller syndrome is becoming an important economic concern since it affects 2 to 3 percent of steers in feedlots. The occurrence of bulling results in physical injury to the steer being mounted, decreased weight gains, and problems in management. In most large feedlot operations separate pens must be established for bullers.

The fact that this mounting is most noticeable in feedlot operations has led investigators to point to crowding and the use of estrogenic implants, such as diethylstilbestrol, as causative factors (Brower and Kiracofe, 1978; Irwin et al., 1979). We should not be misled into believing that markedly reducing the crowding in feedlots and eliminating the use of hormonal implants will eliminate this problem. In the closest living ancestor of cattle, the bison, bulling is fairly common even in free-ranging natural populations. Lott (1983) has noted that males up to the age of five years frequently mount other males, especially those that appear to solicit mounting. Certain males are mounted in particular, just as with bullers in a feedlot. Among the two to four year old bulls such homosexual mounting is commoner than heterosexual mounting. The more mature bison bulls of five years and older rarely engage in homosexual mounting. Feedlot steers fall into the young bull range. The buller syndrome is probably a "naturally" occurring phenomenon that is enhanced by crowding and hormonal implants.

■

SEXUAL DYSFUNCTION

In the wild sexual dysfunction is self-limiting. Males and females that have no interest in the opposite sex, that continuously repel sexual advances, or that do not have the appropriate physiological reactions simply do not reproduce themselves. In the Old MacDonald's farm of bygone years, where all breeding was left to the animals' own designs, the same was true. In fact, husbandry practices tended to select for the most efficiently breeding animals. Hence, it has been said that selective breeding on the domestic scene has enhanced sexual activity over that of the wild ancestors. Under current conditions, however, it is different. We will do anything to breed certain prized animals. Think of the financial implications of dispensing with a famous race stallion because he is a slow or reluctant breeder. Any degree of pampering, coaching, or bribing to get him to breed mares or donate sperm for artificial insemination is worth it. The same goes for valuable livestock animals and pet dogs and cats. I have already alluded to some of the problems with breeding females. Some females may come into physiological estrus but not accept a male's sexual advances. If estrus can be detected by microscopic examination of vaginal cells, these females can be bred artifically. Admittedly we perpetuate the sexual dysfunction, but we get the valuable offspring.

Problems with sexual behavior of breeding males are widespread and quite common. For example, as many as one-third of all breeding rams in some localities have been estimated to be nonbreeders (Zenchak and Anderson, 1980). As with females these problems may be overcome by bypassing nature. If a male has a fear of females, he might be trained to an artificial vagina or subjected to electroejaculation. Bulls that are used for ejaculation into an artificial vagina year in, year out can become jaded and produce poor quality semen. One report documents that periodic stimulation by a real female improves semen quality (Kerruish, 1955).

An overall view of sexual dysfunction, in which we can see several possible etiological factors, is important in domestic animals just as it is in humans. One approach is to define the problem based on where it occurs in the mating sequence. Some males, for example, may show no interest at all in females and never attempt to mount. Others may mount but display an impairment in erection or ejaculating. Analysis of dysfunction on this basis has been particularly useful in stallions (Pickett and Voss, 1975).

A second approach is to differentiate between physiological and behavioral causes of dysfunction. This is the usual approach in dealing with sexual impotence of human males. The first factor one might think of, low testosterone blood concentration, is rarely a physiological factor in animals. Evidence suggests that normally testosterone concentration is much higher than that needed to maintain normal sexual function, except in aging males. Disease, undernutrition, and fatigue may reduce a male's interest in copulation, but these factors should be fairly obvious. Interestingly, restricting the crude protein ration of growing bull calves, so that they were noticeably smaller than well-fed control bulls, did not reduce their sexual performance. In fact, they were more efficient breeders (Wierzbowski, 1978). This is a different matter than severe malnutrition.

We may look for neurochemical reasons for low sexual interest. Some work on rats has implicated excessive amounts of morphinelike chemicals (endorphins) found naturally in the brain as the cause of sexual inactivity. Administration of naloxone, an inhibitor of the endorphin action, led to the onset of sexual activity in sexually inactive rats (Gessa et al., 1979).

A well-documented physiological cause of impotence in animals is abnormality or injury of the penis. In bulls the blood pressure of the penis can reach many times that of systemic blood pressure. Under such pressure the fibrous capsule surrounding the erectile tissue can rupture, making erection impossible (Hudson et al., 1972). Anatomical abnormalities in the venous drainage of the penile erectile tissue can also prevent erection (Ashdown et al., 1979). Physical problems with the erection of the penis can eventually affect sexual interest as the animal experiences repeated inability to copulate.

Behavioral causes of sexual dysfunction are suspect in males when an animal that has been frequently used in stud service suddenly refuses to copulate. The common pattern is for a bull or stallion to mount and appear sexually excited, but then suddenly dismount. Pain, mismanagement, overuse, and rough treatment are common causes of this type of copulatory failure. Pickett and Voss (1975) note that the discipline and rough training methods necessary for racing and showing stallions are not consistent with good breeding management and are probably factors in the high incidence of sexual dysfunction in these animals. Treatment of sexual problems due to rough handling or mismanagement involves using a variety of females and lots of patience. Gradually adapting a stallion to mounting a dummy for artificial collection is one procedure that has proved successful in treating this problem (Kenney and Cooper, 1974).

The you-forgot-what-you-were-missing strategy may also work. In one approach a handler allowed an intromission-shy bull to mount a female—knowing that the animal had a tendency to back off a second or two later. However, just after the bull mounted he was surprised by the slapping on of an artificial vagina over the penis. The bull performed happily ever after (Kendrick, 1954).

One investigator has noted that a stallion refused to breed mares with foals whereas he had readily bred these mares before. When the hindquarters, flanks, abdomen, and udder of the mare were rubbed with the stallion's feces, he then promptly bred the mares. This indicates that olfactory cues of foals may at times inhibit sexual activity (Veeckman, 1979).

Another behavioral contribution to sexual impotence may be developmental. Males that are reared away from females may show no sexual inclination toward females. Some breeders have noted that in dogs, males and females that have been kept apart during several estrous periods no longer will mate. Similarly, boars housed away from sexually receptive females may achieve fewer copulations than those housed near females (Hemsworth et al., 1977). Rams raised alone, but where they could see, hear, and smell other sheep, were all sexually normal as adults. However, about one-half of the rams raised in the all-male groups showed little interest in ewes. In fact, when some of the low-response rams sniffed ewes, they appeared to react adversely to the odor and immediately turned and walked away (Zenchak and Anderson, 1980).

One therapeutic approach to stimulating sexual interest in males with low sexual behavior is to allow them to watch other males copulate. This procedure enhances sexual performance in already sexually active beef cattle and dairy goats (Mader and Price, 1984; Price et al., 1984), and stallions with low or abnormal sexual drive were reported more likely to mate when allowed to watch other stallions mate (Pickett et al., 1977).

Raising males in contact only with members of a different species may result in their displaying sexual interest only to the animals they were raised with. This bonding with another species, sometimes referred to as imprinting, has been observed in swine, cattle, sheep, and goats (Sambraus, 1980; Sambraus and Sambraus, 1975).

We have seen in this chapter that as raising animals becomes more and more artificial, and as we continue to interfere with normal rearing conditions, problems with sexual performance will become more frequent and acute. The more animal breeders rely on artificial insemination, the more normal sexual behavior is going to become irrelevant. Thus, we shall see in our domestic animals an increasingly greater deviation in sexual behavior from that of the wild ancestors. The resulting problems must be approached with an understanding of normal sexual behavior and its determinants.

References

ALMQUIST, J. O. 1973. Effects of sexual preparation on sperm output, semen characteristics and sexual activity of beef bulls with a comparison to dairy bulls. *J. Anim. Sci.* 36:331–336.

126   SEXUAL BEHAVIOR OF FEMALES AND MALES

ALMQUIST, J. O., AND E. B. HALE. 1956. An approach to the measurement of sexual behavior and semen production of dairy bulls. *Third Intl. Cong. Anim. Reprod.*, 50–59.

ARON, C. 1979. Mechanism of control of the reproductive function by olfactory stimuli in female mammals. *Physiol. Rev.* 59:229–284.

ARONSON, L. R., AND M. L. COOPER. 1974. Olfactory deprivation and mating behavior in sexually experienced male cats. *Behav. Biol.* 11:459–480.

ASHDOWN, R. R., J. S. E. DAVIS, AND C. GIBBS. 1979. Impotence in the bull: Abnormal venous drainage of the corpus cavernosum penis. *Vet. Rec.* 104:423–428.

BACK, D. G., B. W. PICKETT, J. L. VOSS, AND G. E. SEIDEL. 1974. Observations on the sexual behavior of nonlactating mares. *J. Amer. Vet. Med. Assn.* 165:717–720.

BANKS, E. M. 1964. Some aspects of sexual behaviour in domestic sheep, *Ovis aries. Behaviour* 23:249–279.

BARFIELD, R. J., AND L. A. GEYER. 1972. Sexual behavior: Ultrasonic post-ejaculatory song of the male rat. *Science* 1976:1348–1350.

BEACH, F. A. 1976. Sexual attractivity, proceptivity, and receptivity in female mammals. *Horm. Behav.* 7:105–138.

BEACH, F. A., AND B. J. LE BOEUF. 1967. Coital behaviour in dogs. I. Preferential mating in the bitch. *Anim. Behav.* 15:546–580.

BEAMER, W., G. BERMANT, AND M. T. CLEGG. 1969. Copulatory behaviour of the ram (*Ovis aries*). II. Factors affecting copulatory satiation. *Anim. Behav.* 17:706–711.

BERMANT, G., M. T. CLEGG, AND W. BEAMER. 1969. Copulatory behaviour of the ram (*Orvis aris*). I. A normative study. *Anim. Behav.* 17:700–705.

BLOCKEY, M. A. DE B. 1978. The influence of serving capacity of bulls on herd fertility. *J. Anim. Sci.* 46:589–595.

BLOCKEY, M. A. DE B. 1981a. Development of serving capacity test for beef bulls. *Appl. Anim. Ethol.* 7:307–319.

BLOCKEY, M. A. DE B. 1981b. Modification of a serving capacity test for beef bulls. *Appl. Anim. Ethol.* 7:321–336.

BLOCKEY, M. A. DE B. 1981c. Further studies on the serving capacity test for beef bulls. *Appl. Anim. Ethol.* 7:337–350.

BOOTH, W. D., AND B. A. BALDWIN. 1980. Lack of effect on sexual behavior or the development of testicular function after removal of olfactory bulbs in prepubertal boars. *J. Reprod. Fert.* 58:173–182.

BOUCHER, J. H., R. H. FOOTE, AND R. W. KIRK. 1957. The evaluation of semen quality in the dog and the effects of frequency of ejaculation upon semen quality, libido, and depletion of sperm reserves. *Cornell Vet.* 48:67–86.

BROWER, G. R. AND G. H. KIRACOFE. 1978. Factors associated with the buller-steer syndrome. *J. Anim. Sci.* 46:26–31.

DIAMOND, M., A. L. DIAMOND, AND M. MAST. 1972. Visual sensitivity and sexual arousal levels during the menstrual cycle. *J. Nerv. Ment. Dis.* 155:170–176.

EMERY, D. E., AND B. D. SACHS. 1975. Ejaculatory pattern in female rats without androgen treatment. *Science* 190:484–486.

ESSLEMONT, R. J., AND M. BRYANT. 1976. Oestrous behavior in a herd of dairy cows. *Vet. Rec.* 99:472–475.

ESSLEMONT, R. J., R. G. GLENCROSS, M. J. BRYANT, AND G. S. POPE. 1980. A quantitative study of pre-ovulatory behavior in cattle (British Friesian heifers). *Appl. Anim. Ethol.* 6:1–17.

FABRE, C. 1977. Existence of an ejaculatory-like reaction in ewes ovariectomized and treated with androgens in adulthood. *Horm. Behav.* 9:150–155.

FLETCHER, I. C., AND D. R. LINDSAY. 1968. Sensory involvement in the mating behaviour of domestic sheep. *Anim. Behav.* 16:410–414.

FOOTE, R. H. 1975. Estrus detection and estrus detection aids. *J. Dairy Science* 58:248–256.

GESSA, G. L., E. PAGLIETTI, AND B. PELLEGRINI QUARAMTOTTI. 1979. Induction of copulatory behavior in sexually inactive rats by naloxone. *Science* 204:302–304.

GEYER, L. A., R. J. BARFIELD, AND T. K. MC INTOSH. 1978a. Influence of gonadal hormones and sexual behavior on ultrasonic vocalization in rats. II. Treatment of Males. *J. Comp. Physiol. Psychol.* 92:447–456.

GEYER, L. A., T. K. MC INTOSH, AND R. J. BARFIELD. 1978b. Effects of ultrasonic vocalizations and male's urine on female readiness to mate. *J. Comp. Physiol. Psychol.* 92:457–462.

GOODWIN, M., K. M. GOODING, AND F. REGNIER. 1979. Sex pheromone in the dog. *Science* 203:559–561.

GRANDAGE, J. 1972. The erect dog penis: A paradox of flexible rigidity. *Vet. Rec.* 91:141–147.

GRUBB, P., AND P. A. JEWELL. 1973. The rut and occurrence of oestrus in the Soay sheep on St. Kits. *J. Reprod. Fert. Suppl.* 19:491–502.

HAMILTON, W. D. 1971. Geometry for the selfish herd. *J. Theor. Biol.* 31:295–311.

HART, B. L. 1970. Mating behavior in the female dog and the effects of estrogen on sexual reflexes. *Horm Behav.* 2:93–104.

HART, B. L. 1978*a*. Hormones, spinal reflexes and sexual behavior. In *Biological determinants of sexual behavior*, pp. 205–242, ed. J. B. Hutchinson. New York: Wiley.

HART, B. L. 1978*b*. Reflexive aspects of copulatory behavior. In *Sex and behavior: Status and prospectus*, pp. 205–242, eds. T. E. McGill, D. A. Dewsbury, and B. Sachs. New York: Plenum.

HART, B. L. 1983. Flehmen behavior and vomeronasal organ function. In *Chemical signals in vertebrates III*, pp. 87–103, eds. R. M. Silverstein and D. Muller-Schwarze. New York: Plenum.

HART, B. L., AND C. M. HAUGEN. 1971. Scent marking and sexual behavior maintained in anosmic male dogs. *Comm. Behav. Biol.* 6:131–135.

HEMSWORTH, P. H., R. G. BEILHARZ, AND D. B. GALLOWAY. 1977. Influence of social conditions during rearing on sexual behavior of domestic boar. *Anim. Prod.* 24:245–251.

HUDSON, R. S., S. D. BECKETT, AND D. F. WALKER. 1972. Pathophysiology of impotence in the bull. *J. Amer. Vet. Med. Assoc.* 161:1345–1347.

HULET, C. V., S. K. ERCANBRACK, R. L. BLACKWELL, D. A. PRICE, AND L. O. WILSON. 1962*a*. Mating behavior of the ram in the multi-sire pen. *J. Anim. Sci.* 21:865–869.

HULET, C. V., R. L. BLACKWELL, S. K. ERCANBRACK, D. A. PRICE, AND L. O. WILSON. 1962*b*. Mating behavior of the ewe. *J. Anim. Sci.* 21:870–874.

HURNIK, J. F., G. J. KING, AND H. A. ROBERTSON. 1975. Estrus and related behaviour in postparturient Holstein cows. *Appl. Anim. Ethol.* 2:55–68.

IRWIN, M. R., D. R. MELENDY, M. S. AMOSS, AND D. P. HUTCHESON. 1979. Roles of predisposing factors and gonadal hormones in the buller syndrome of feedlot steers. *J. Amer. Vet. Med. Assn.* 174:367–370.

JENNINGS, J. J., AND J. MC WEENEY. 1976. Effect of frequent ejaculation on semen characteristics in rams. *Vet. Rec.* 98:230–233.

JÖCHLE, W. 1975. Current research in coitus-induced ovulation: A review. *J. Reprod. Fert. Suppl.* 22:165–207.

JORGENSON, J. W., M. NOVOFNY, M. CARMACK, G. B. COPLAND, S. R. WILSON, W. K. WHITTEN, AND S. KATONA. 1978. Chemical scent constituents in the urine of the red fox (*Vulpes vulpes* L.) during the winter season. *Science* 199:796–797.

KENDRICK, J. 1954. Psychic impotence in bulls. *Cornell Vet.* 1054:289–293.

KENNEY, R. M., AND W. L. COOPER. 1974. Therapeutic use of a phantom for semen collection from a stallion. *J. Amer. Vet. Med. Assn.* 165:706–707.

KERRUISH, B. M. 1955. The effect of sexual stimulation prior to service on the behaviour and conception rate of bulls. *Brit. J. Anim. Behav.* 3:125–130.

KIDDY, C. A., D. S. MITCHELL, D. J. BOLT, AND H. W. HAWK. 1978. Detection of estrus-related odors in cows by trained dogs. *Biol. Reprod.* 19:389–395.

KLEIMAN, D. 1977. Monogamy in mammals. *Quart. Rev. Biol.* 52:39–69.

LADEWIG, J., AND B. L. HART. 1980. Flehmen and vomeronasal organ function in male goats. *Physiol. Behav.* 24:1067–1071.

LADEWIG, J., AND B. L. HART. 1981. Demonstration of estrus-related odors in cow urine by operant conditioning of rats. *Biol. Reprod.* 24:1165–1169.

LADEWIG, J., E. O. PRICE, AND B. L. HART. 1980. Flehmen in male goats: Role in sexual behavior. *Behav. Neural Biol.* 30:312–322.

LESMEISTER, J., AND E. ELLINGTON. 1977. Effect of steroid implants on sexual behavior of beef calves. *Horm. Behav.* 9:276–280.

LINDSAY, D. R., AND I. C. FLETCHER. 1972. Ram-seeking activity associated with oestrous behaviour in ewes. *Anim. Behav.* 20:452–456.

LOTT, D. F. 1983. The buller syndrome in American bison bulls. *Appl. Anim. Ethol.* 11:183–186.

MAC LEOD, N., W. REINHARDT, AND F. ELLENDORF. 1979. Olfactory bulb neurons of the pig respond to an identified steroidal pheromone and testosterone. *Brain Res.* 164:323–327.

MADER, D. R., AND E. O. PRICE. 1984. The effects of sexual stimulation on the sexual performance of Hereford bulls. *J. Anim Sci.* 59:294–300.

MELROSE, D. R., H. C. B. REED, AND R. L. S. PATTERSON. 1971. Androgen steroids associated with boar odour as an aid to the detection of oestrus in pig artificial insemination. *Brit. Vet. J.* 127:495–501.

MICHAEL, R. P. 1963. Observations upon the sexual behaviour of the domestic cat (*Felis catus* L.) under laboratory conditions. *Behaviour* 18:1–24.

MICHAEL, R., AND D. ZUMPE. 1978. Potency in male rhesus monkeys: Effects of continuously receptive females. *Science* 200:451–452.

MORRIS, N. M., AND J. R. UDRY. 1970. Variations in pedometer activity during the menstrual cycle. *Obstet. Gynec.* 35:199–201.

MYLREA, P. J., AND R. G. BEILHARZ. 1964. The manifestation and detection of oestrous in heifers. *Anim. Behav.* 12:25–30.

O'CONNELL, R. J., A. G. SINGER, F. MACRIDES, C. PFAFFMANN, AND W. C. AGOSTA. 1978. Responses of the male golden hamster to mixtures of odorants identified from vaginal discharge. *Behav. Biol.* 24:244–255.

PEPELKO, W. E., AND M. T. CLEGG. 1965. Studies of mating behaviour and some factors influencing the sexual response in the male sheep *Ovis aries. Anim. Behav.* 13:249–258.

PICKETT, R. W., AND J. L. VOSS. 1975. Abnormalities of mating behaviour in domestic stallions. *J. Reprod. Fert. Suppl.* 23:129–134.

PICKETT, B. W., J. L. VOSS, AND E. I. SQUIRES. 1977. Impotence and abnormal sexual behavior in the stallion. *Theriogeneology* 8:329.

PRICE, E. O., V. M. SMITH, AND L. S. KATZ. 1984. Sexual stimulation of male dairy goats. *Appl. Anim. Behav. Sci.* 13:83–92.

PRICE, E. O. (In press.) Sexual behavior of large domestic farm animals: An overview. *J. Anim. Sci.*

RANDEL, R. D., R. E. SHORT, D. S. CHRISTENSEN, AND R. A. BELLOWS. 1975. Effect of clitoral massage after artificial insemination on conception in the bovine. *J. Anim. Sci.* 40:1119–1123.

SAMBRAUS, H. H. 1980. Imprinting in farm animals: Implications for sexual behavior. *Bovine Pract.* 1(4):8–9.

SAMBRAUS, H. H., AND D. SAMBRAUS. 1975. Prägung von Nutztieren auf Menschen. *Z. Tierpsychol.* 38:1–17.

SCHELS, H. F., H. ANSARI, AND D. MOSTAFAWI. 1978. The use of heat mount detectors in a large Iranian dairy herd. *Vet. Rec.* 102:211–212.

SIGNORET, J. P. 1970. Reproductive behaviour of pigs. *J. Repro. Fert. Suppl.* 11:105–117.

SIGNORET, J. P. 1971. The reproductive behaviour of pigs in relation to fertility. *Vet. Rec.* 88:34–38.

SIGNORET, J. P. 1975. Influence of the sexual receptivity of a teaser ewe on the mating preference in the ram. *Appl. Anim. Ethol.* 1:229–232.

SINGER, A. G., W. C. AGOSTA, R. J. O'CONNELL, C. PFAFFMAN, D. V. BOWEN, AND F. H. FIELD. 1976. Dimethyl disulfide: An attractant pheromone in hamster vaginal secretion. *Science* 191:948–950.

STEBENFELDT, G., AND V. M. SHILLE. 1977. Reproduction in the dog and cat. In *Reproduction in domestic animals*, 3rd ed., pp. 499–527, eds. H. H. Cole and P. T. Cupps. New York: Academic.

THIERY, J. C. AND J. P. SIGNORET. 1978. Effect of changing the teaser ewe on the sexual activity of the ram. *Appl. Anim. Ethol.* 4:89–90.

VAN DEMARK, N. L., AND R. L. HAYS. 1952. Uterine motility responses to mating. *Amer. J. Physiol.* 170:518–521.

VEECKMAN, J. 1979. Aberrant sexual behaviour of a covering stallion and its ethological solution. *Appl. Anim. Ethol.* 5:292.

VEECKMAN, J., AND F. O. ODBERG. 1978. Preliminary studies on the behavioural detection of oestrus in Belgian "warm-blood" mares with acoustic and tactile stimuli. *Appl. Anim. Ethol.* 4:109–118.

WIERZBOWSKI, S. 1978. The sexual behavior of experimentally underfed bulls. *Appl. Anim. Ethol.* 4:55–60.

WILLIAMSON, N. B., R. S. MORRIS, D. C. BLOOD, C. M. CANNON, AND P. J. WRIGHT. 1972. A study of oestrous behaviour and oestrus detection methods in a large commercial diary herd. *Vet. Rec.* 91:58–62.

WYSOCKI, C. J. 1979. Neurobehavioral evidence for the involvement of the vomeronasal system in mammalian reproduction. *Neurosci. Biobehav. Rev.* 38:301–341.

ZENCHAK, J. J., AND G. C. ANDERSON. 1980. Sexual performance levels of rams (*Ovis aries*) as affected by social experiences during rearing. *J. Anim. Sci.* 50:167–174.

CHAPTER FOUR ■ MATERNAL BEHAVIOR AND MOTHER-YOUNG INTERACTIONS

One of the marvels of animal behavior is that a new mother, with no previous opportunity to observe or engage in care of young, suddenly performs a relatively complex set of maternal tasks that continue and change until the offspring are able to survive on their own. Not only are various elements of her maternal role performed with perfect timing and precision, but she has an emotional attachment to her young, representing the strongest interindividual bond in nature. The mother-young bond is the only instance in nature where an animal is so willing to risk its life for another animal.

In dairy cattle and goats, young are normally removed from the dam at birth or very soon thereafter so that the maximum amount of milk is available for commercial sale. Hence, maternal care of the offspring is of little concern with these animals. In the other domestic animals, the performance of maternal behavior affects the health and growth rate of the offspring, and maternal care is a form of farm labor. Although we can step in and provide the milk and warmth that a mother normally provides, the natural mother can do a better job and save us considerable time and expense.

In young animals that are destined to be pets or breeding stock, we have particular interest in their behavior. Interactions with the mother provide the young with the basis for normal social and sexual behavior. As described in Chapter 7, curtailing the rich mother-young interactions by separating young from their mothers can lead to major behavioral disturbances in the offspring as they mature.

■

GENERAL CHARACTERISTICS OF MATERNAL BEHAVIOR

We will find it conceptually most easy to discuss maternal behavior by separating domestic animals into those species that are monotocous, that is, that give birth typically to just one young, and those species that are polytocous species, that give birth to an entire litter of young. The control of maternal responsiveness, the reproductive strategies, and our management of mothers and their offspring are influenced by major behavioral differences between the monotocous and polytocous species. Therefore, before looking at the two categories of maternal behavioral patterns, it is instructive to consider first the induction of maternal responsiveness, reproductive strategies of motherhood in general, and how behavior of domestic mothers compares to wild mothers of related or ancestral species.

Induction of Maternal Responsiveness

The strong attachment of mothers for their newborn infants, as well as maternal care-giving tasks, are activated at the time of birth partially by hormonal changes. The specific hormonal conditions that produce maternal responsiveness are not understood even in laboratory animals. Thus, unlike the ease with which sexual behavior may be induced with hormone injections, it is still impossible to produce completely normal behavior in virgin females by injecting them with exogenous hormones. There is evidence that the main function of hormonal changes at the time of parturition is to activate maternal behavior, and that stimuli provided by the young maintain the behavior. Hormonal changes are not essential for maternal responsiveness, for without any apparent hormonal induction, retrieving, nest building, and care of the young can be induced in virgin female rats, and even male rats, by exposing them continuously to young rat pups for from 10 to 15 days (Rosenblatt, 1967).

An example of the role of stimuli from newborn in maintaining maternal behavior is illustrated by a study on rats showing that the stimulation provided by just a single pup is not enough to maintain maternal behavior in a lactating female (Leigh and Hofer, 1973). If this is true of other species giving birth to more than one young at a time, such as swine, dogs, and cats, the experiment suggests that our assessment of maternal responsiveness in a female who has just one offspring may not be representative of her maternal responsiveness with a large litter. The observation also indicates that one should carefully observe dog and cat mothers of single-offspring litters for signs of inadequate maternal care and be prepared to provide auxiliary maternal care. If the litter is larger the next time, the mother may be stimulated to display a greater degree of maternal care.

Reproductive Strategies

I have alluded to the fact that there are hormonal changes in females approaching and going through parturition that are responsible for the induction of maternal behavior. A hormonal change of a different type also occurs that appears to be related to the fact that lactation and constant maternal care are vital to the welfare of the young, but extremely demanding on the mother's resources. Typically if an animal is upset by environmental disturbances, secretion of corticosteroids from the adrenal gland causes mobilization of energy stores in preparation for an emergency. In lactating mothers, however, such a response would place demands on resources that are also necessary for the young. In laboratory rats, and presumably in other species, the lactating female is buffered against disturbing physical and emotional stimuli in the sense that such stimuli produce a smaller adrenal secretory response than in nonlactating females (Stern and Levine, 1972). The idea is that rather than preparing to protect herself by utilization of energy sources and general systemic activation, the mother takes threatening stimuli rather calmly, which, of course, puts her own long-term safety somewhat in jeopardy.

We must remember that mother love in animals is only one aspect of reproductive strategy. Parental care is vital to reproductive success, but maximum reproductive success dictates that the mother must stop contributing her nutritional resources to current young as soon as they are capable of surviving on their own. Each day of nursing a rapidly growing offspring saps nutritional resources from a mother that could go toward building up her body nutritional stores for her next pregnancy and lactation. Cost-benefit analysis dictates a mother will want to wean her young as soon as it is capable of doing well on its own. A cost-benefit analysis approached from the offspring's standpoint dictates the young will attempt to take the nutrient-rich milk of its mother as long as possible, even after it is fully capable of foraging on its own. What we see, therefore, is a mother-young conflict, with the mother rejecting the young at the time of weaning and the young still attempting to suckle before being pushed away by the mother.

The natural weaning process is something we do not often see on farms, because young animals are separated from their mother when we feel the young can survive on their own. When the young are left with their mothers and natural weaning allowed, the bond between mothers and young can last beyond the time of weaning. Lambs, for example, will continue to maintain contact with their ewes when grazing or resting for some time after weaning is completed (Arnold and Pahl, 1974). If ewes are receiving a restricted diet, the time of weaning will occur sooner, and the attachment of the mother to the young diminishes sooner (Arnold et al., 1979).

Domestic Mothers Compared to Wild Mothers

The survival of young animals in the domestic environment is usually ensured by animal keepers. With pet dogs, cats, and horses, we do not hesitate to intervene just for the comfort of the mother or young. During parturition, owners may help by assisting in the birth process, cleaning and grooming the newborn, stimulating respiration, and helping the newborn to locate a nipple. Assisting in the birth of calves and removing retained placentas are frequent in cattle farming, and helping lambs to nurse or find their mothers are a part of lambing operations.

The outcome of centuries of this type of intervention in parturition and care of neonates has been the removal of natural selection pressures that maintained the pattern of maternal behavior within fairly narrow limits in the wild. Under natural selection pressures, females not providing optimal maternal care simply did not reproduce themselves. Once the forces acting to maintain maternal responses within narrow limits were eliminated by husbandry efforts, the genetic basis of maternal behavior became quite variable. Thus, some females display a remarkable orchestration of optimal maternal responses, and others show practically no interest in their offspring; there are, of course, gradations in between.

When females of various domestic species have attempted to make it in the wild without human assistance, only a few have been able to reproduce themselves, but those that do are excellent mothers. All domestic species have their feral counterparts, which are animals that have escaped to the wild or been turned loose intentionally. Cats seem to adapt most easily to a feral existence, undoubtedly because we have historically intervened the least in the care of kittens and adult cats.

It was observed that sheep allowed to go feral in New Zealand initially declined in numbers, not because of problems with adults surviving, but because of inadequate maternal care and protection of the lambs. For example, females might not seek a sheltered enough location in which to give birth to the young. Those females that did give sufficient maternal care continued to reproduce themselves, and their offspring eventually supplied the breeding stock of the group. Within a few generations, the sheep appeared to be maintaining their numbers almost as well as their captive counterparts on adjacent farms under the care of a farmer.

General Comments on Maternal Behavior

Differences in maternal behavior among domestic mammals are closely related to the number of offspring born at one time and the state of maturation of the young at birth. Monotocous species are those that ordinarily give birth to just one or two young, such as horses, cattle, and sheep. The young are relatively mature, or precocial, at birth. Their eyes and ears are functioning fairly well, and they are capable of running after their mothers within a few hours after birth.

Species that are classified as polytocous are those that ordinarily give birth to more than one or two young, including dogs, cats, and swine. The young of dogs and cats are relatively immature, or altricial, at birth. For example, the eyes and ears of dogs and cats do not function until the second or third week of life, and their locomotor abilities are restricted to those needed to crawl around to find a nipple. In terms of the polytocous-monotocous classification, swine are intermediate because, although several young are born in a litter, the young are quite precocial. Primates, including humans, are classified as monotocous, but the young are immature.

In most monotocous species with precocial young, there is a very close bond between the young and the mother, and the mothers actively reject all alien young very soon after parturition. In the polytocous species, there is no such individual bonding, and the mother will usually adopt alien young. There are a number of classic examples where mother dogs and cats have adopted a young animal even of a different species.

Differences in behavior at the time of parturition between primiparous females (giving birth for the first time) and multiparous females (having given birth at least once previously) are relatively minor. Experienced mothers appear to respond somewhat more readily to the neonates in licking and grooming and are less disturbed by the physiological changes during birth. Although some females may be calmer and more easily handled while lactating, there is no evidence suggesting that permanent behavioral effects stem from the experience of parturition and care of offspring.

■

MATERNAL BEHAVIOR OF MONOTOCOUS SPECIES

The wild ancestors of cattle, sheep, goats, and horses lived in large flocks or herds. Many maternal responses can be understood in terms of the necessity of the females protecting the newborn from predation by staying with the herd, and the necessity, also, of forming a bond with their own offspring so as not to expend effort raising unrelated offspring of the herd. In the wild it is characteristic for a female about to give birth to briefly separate from the herd. With no other offspring around, she bonds to her own newborn. Once having bonded to her newborn the mother then returns to the group where she will administer maternal care only to her own offspring and vigorously reject all others. Separation from the herd poses some chance of predation. However, the long-term reproductive advantage of the mother's being able to bond to her own offspring has such a strong genetic payoff that the behavior has been maintained by natural selection (Lott, 1973). Although a young animal learns to recognize its mother, it does not form an attachment to its mother so strong that it will not suckle from other lactating females.

Through the process of domestication, the behavior of rejecting alien young has diminished, but the rejection behavior is still very pronounced in sheep and goats. This is one aspect of natural maternal behavior that is a disadvantage in the domestic environment since it makes it difficult to provide a foster mother for orphaned animals.

Behavior before and during Parturition

For the greater part of the duration of gestation there are few signs that distinguish the pregnant female from other females. When the time of parturition is 24 to 48 hours away the female will often show signs of restlessness, and as the time of parturition approaches, there are some signs of pain as evidenced by the tendency of some females to kick at the flank or roll. This is especially apparent in the mare. Pregnant females about to give birth are often separated from the herd or flock by the farmer or rancher, so that the tendency of a female to separate on her own from the herd may not be apparent.

There are three stages of labor that constitute parturition. The first stage, contraction, is when the animal does a great deal of straining. This is sometimes accompanied by urination and defecation. The straining is related to the movement of the fetus from the uterus into the birth canal. Usually, the female will be lying down during most of parturition; however, during the first stage of labor many females may alternate between lying down and standing. Toward the end of this stage the first water bag, or the allantochorion, is broken, releasing a great deal of straw-colored fluid.

The second stage of labor, delivery, is when the fetus moves through the birth canal. The most intense straining occurs during this stage. The second placental membrane, the amnion, first appears at the external genitalia; it may be broken or intact. The delivery of the fetus follows next.

In a normal birth the head is the first part of the fetus to move through the birth canal. While the head of the fetus is moving, the female is almost always lying down. Most of the straining is involved in moving the head of the fetus. After the head is out, cows and sheep may stand and dangle the head, front legs, and shoulder of the newborn. As a result of this lifting movement, the fetus frequently drops through the birth canal. A common emergency in cattle and horses, to a lesser extent in sheep and goats, is abnormal presentation of the fetus requiring medical assistance.

The third stage, delivery of the placenta, immediately follows the second stage. The placenta is usually delivered 30 minutes to six hours following the birth of the newborn. There is some straining associated with expulsion of the placenta, but much less than in the second stage. A retained placenta is one of the commoner medical problems in cattle reproduction. Immediately after the birth of the newborn the mother invariably licks the birth fluids and pulls remaining placental membranes from the newborn. If this is not done, the neonate may die. The placenta is commonly eaten by the cow and ewe soon after it is expelled.

Function of Placentophagy The attraction to the placenta may be so pronounced that a mother resists the taking away of the placenta more than her own young. Placentophagy is almost, but not quite, universal among mammalian species. The behavior appears to be absent in all human cultures, and it does not occur in aquatic mammals or marsupials. Among domestic animals it is doubtful if horses or camels often eat the placenta (Kristal, 1980). Of course, animals may not have the opportunity to eat the placenta if we remove it first or a scavenger comes along and takes it.

Are the causes or consequences of placentophagy the same in herbivores as carnivores? In nesting carnivores such as dogs and cats, we can see the advantage of placentophagy in keeping the nest clean and free of material in which bacteria might thrive. Furthermore, the placenta offers some nutritive value to the mother in the form of protein and water, allowing her to stay in the nest longer with the newborn before leaving to feed.

In the large herbivores, especially those that stay around the place of birth for some time, or those that hide their newborn, we can see the advantage in removing a source of attraction of predators. Some predators are able to locate a mother and new-born by detecting the activity of avian scavengers such as vultures that are circling and awaiting the expulsion of placentas (Estes, 1967; Kruuk, 1972; Schaller, 1972). Thus, by promptly consuming the placenta, herbivores are able to avoid attracting vultures, which in turn are noticed by predators some distance away.

These evolutionary hypotheses suggest that when it is not especially advantageous in terms of offspring survival for mothers to eat the placenta they should not. Detailed observations on the wildebeest point along these lines (Estes and Estes, 1979). The newborn wildebeest gains locomotor coordination more quickly after birth than any other ungulate, and wildebeest mothers move away from the birth area within a few minutes after birth. Wildebeest cows rarely eat afterbirths, even if they stay around the birth site for a long time and are not harassed by predators or scavengers. Newborn equids are about as precocial as wildebeests in being able to stand and run with the mother, and in feral horses, placentophagy appears to be uncommon.

These ultimate explanations of placentophagy do not address the immediate causes of the behavior. We can only assume that placentophagic females develop a strong desire for the placenta. It would be illogical to assume they are aware of the consequences of eating it.

There are additional biological roles that placentophagy may take. The placenta is a large, complex organ that carries on secretory, filtration, and transport functions in situ. When expelled, it is full of hormones, including estrogen, progesterone, and prolactin. No one has found a difference in maternal responsiveness that corresponds to whether mothers are allowed to consume placentas or not, but there is some evidence that hormones may contribute to the production or delivery of milk in rats and cows (Kristal, 1980; Zarrow et al., 1972).

Another possible consequence of placentophagy that deserves serious attention is that of the effects of the placenta on the maternal immune-response system. Kristal (1980) has proposed that the placenta contains factors that prevent a mother from forming antibodies against fetal antigens, which might impair subsequent pregnancies (he cites the example of RH-incompatibility). Even the estrogen and progesterone present in the placenta may suppress such immunological consequences.

One of the problems with this hypothesis, as with the hormonal hypothesis, is that the health and behavior of mother and young appear to be unaffected by preventing the mother from eating the placenta. It could be, as suggested by some, that placentophagy is physiologically significant only when other conditions for infant survival are marginal (Zarrow et al., 1972). In terms of an overall understanding of placentophagy, Kristal notes that perhaps there are several advantageous consequences of placentophagy and that the physiology and ecological environment of each species dictates which of the many possible benefits is the most important.

Postparturient Behavior

Within an hour or two after birth most of the newborn make active attempts to suckle. Many times this appears to involve a great deal of trial-and-error searching that, to the human observer, appears extremely frustrating. After a few successful attempts at finding the teat most young seem to learn the appropriate responses very rapidly, and mothers may be instrumental in guiding young to the teat. Nursing can be expected to begin within two to five hours after birth. There may be some olfactory cues that help to guide the newborn to the teat, but this has not been examined in monotocous species.

The initial nursing attempts are independent of the newborn's need for milk. That is, the young animals are born with an inherent tendency to nurse before they become hungry. The adaptive value of this separation of nursing behavior from hunger is obvious. Since the first nursing attempts are more or less trial and error, it is important that a newborn animal begin the task before it is too hungry, lest it become too weak before it has mastered the technique.

Although one would not expect the visual acuity of a newborn to be very useful for locating a nipple soon after birth, the newborn does have some visual capabilities in at least locating its mother and the mammary gland. Some investigators have observed that a newborn has a tendency to move toward large objects, particularly ones with which some physical contact has already been made. The mother often facilitates the approach of its young to the udder by her tendency to lick it. As the young approaches her, head first, the mother licks the head region, then the neck, then the shoulder and back. As she does this,

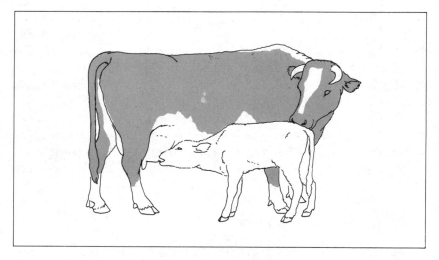

Figure 4-1 When a mother cow licks a newborn, she tends to bring her udder into close proximity to the newborn's head.

she automatically brings her udder into closer proximity to the newborn's head (Figure 4-1).

Licking of newborn goats and sheep by their mother may be so vigorous that the newborn are often knocked off their feet. It has been noted that when a goat mother has twins, she usually directs most of her licking toward the more vigorous twin. Because the licking actually slows the progress of the more vigorous kid, when the weaker kid does finally struggle to its feet, it is less impeded in its nursing attempts and is likely to nurse first. Thus, the mother acts as a "compensatory mechanism" when kids differ in vigor at birth. Under harsh environmental conditions where the kids are weakened, only the more vigorous kids may gain access to the udder (Klopfer and Klopfer, 1977).

In the monotocous domestic species the young are capable of straying from their mother very shortly after birth. Hence, it is very adaptive that a strong bond of the mother to her young develop to the point where alien young are vigorously rejected. Were it not for the bond of the mother to her young, she would find herself providing nourishment and risking her life in the face of predation for the welfare of another female's offspring. The usefulness of this bond is enhanced if a mother can recognize her young at some distance.

The young have little to lose by seeking nourishment or protection from any mother that will take them. Hence they seem not to have a bond to their mother analogous to her bond to them. But since most mothers will reject them, it behooves the young to readily recognize their own mothers so that they can find them without delay.

Three types of sensory cues used in recognition are, olfactory, auditory, and visual. The olfactory-gustatory system is the most important in close-up recognition of young by a mother, but there is little evidence that a young animal uses olfaction to identify its own mother. At a distance the unique visual and auditory characteristics seem to be the means by which mothers and young recognize each other.

As young animals wander off more and more from their mothers, visual and auditory cues become more important. Interestingly, new mothers seem to be innately attracted to the distress calls of all young. The vocalizations of goat kids do not take on unique individual characteristics until about the fourth day of life (Lenhardt, 1977). Assuming this is true of other species as well, we could say that a mother plays it safe and responds to all young up until the time that she can recognize her own.

Visual cues can be quite important in allowing a mother to screen out others that might approach her. In sheep, where this topic has been studied most extensively, visual cues, if pronounced enough, can even override auditory or olfactory cues (Alexander, 1977; Alexander and Shillito, 1977). When a ewe does not like the visual appearance of a lamb approaching her, she may avoid it by dodging or butt it away irrespective of olfactory and auditory cues from the lamb. For example, if a ewe's normally white lamb is made all black, she will avoid it or even butt it on first approach. To determine the part of the lamb's body the ewe visually attends to, one experiment was designed in which lambs were blackened over various selected parts of their bodies. After a four-hour separation from their mothers they were allowed to run toward their mothers, and the behavior of the ewes in avoiding or dodging their lambs was observed (Figure 4-2). The results indicate that ewes attended mostly to the appearance of the head. A lamb with a head that was all blackened was treated the same as a lamb completely blackened, and a black eye was almost as noticeable to a ewe as a black head.

Lambs begin using visual and auditory cues to identify their mothers after they are a few days old. When prevented from using visual cues, many lambs are able to find their mothers by

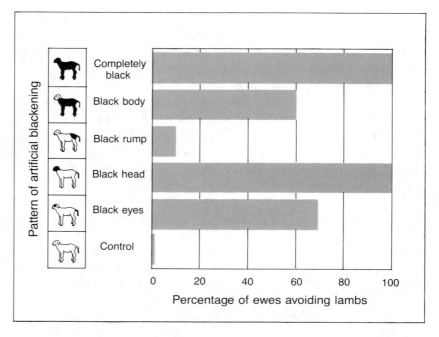

Figure 4-2 Changing the visual appearance of lambs interferes with the mothers' ability to identify their own lambs, and they reject them. In this experiment just blackening the head or the area around the eye had a much greater effect than blackening a comparable amount of skin in the hindquarters. This shows that ewes attend mostly to the head region of their lambs. (Data from Alexander and Shillito, 1977.)

identifying their vocalizations. This ability improves as the lambs get older up to about four weeks of age (Shillito, 1975).

The visual appearance of a ewe, if altered enough, can confuse a lamb in the recognition of its mother. Farmers know that sometimes if a ewe is sheared, her lamb may have difficulty recognizing her. Whether a lamb is born to a shorn or a fleeced ewe, it gets used to its mother's hair style. When confronted with alien ewes, the lambs have a greater tendency to approach ewes that are similar to their own mothers in hair style (Alexander, 1977; Alexander and Shillito-Walser, 1978).

Our information about the involvement of visual and auditory cues in the recognition between mother and young of other domestic monotocous species is limited. Observations of horses indicate that mares are able to recognize their foals at some distance. As with sheep, disturbance of the visual cues interferes with the ability of the mare and foal to recognize each other. Vocalizations, especially the neigh, appear to be non-specific alerting cues, although there is some evidence that mares can recognize neighs of their own foals. When a foal approaches its mother, both exchange low-frequency, low-amplitude, pulsated sounds known as nickers. Whether mares and foals can identify each other's nickers is not known (Wolski et al., 1980).

In the wild ancestors and relatives of the domestic species, visual, auditory, and olfactory cues are all necessary to keep mother and young in contact. Disturbance of visual and auditory cues severely impairs the ability of mothers and offspring to find each other (Arnold et al., 1975; Morgan et al., 1975). The acid test of acceptance, however, involves olfaction. Mothers of all domestic species invariably smell their young on reunion. This is the moment of truth; visual and auditory cues can at times be misleading, but the olfactory bond is almost foolproof.

The development of maternal responsiveness to a specific newborn animal is dependent on contact between the mother and the young within the first few hours postpartum. For example, in goats a five-minute contact is sufficient to establish a bond between an inexperienced mother and young, but acceptance may not occur if a one- or two-hour separation precedes contact (Klopfer, 1971; Lickliter, 1982). If no contact is allowed between dairy cattle mothers and young for five hours after birth, a maternal bond is not formed in half of the animals (Hudson and Mullord, 1977). These findings contrast with those in sheep, in which complete acceptance occurs for as long as six hours postpartum, even when lambs and dams are separated at birth without initial contact (Smith et al., 1966).

Like other natural laws of animal behavior, the law of the bond is broken under domestic conditions. With sheep, new mothers or pregnant females may steal lambs from other mothers. Such mismothering probably reflects intensive husbandry conditions that make it impossible for a mother to temporarily withdraw from the herd or flock and bond with her own young. In the emotionally charged situation in which a ewe is near parturition, her proximity to other ewes that have just lambed and her attraction to the birth fluids may create real confusion in the bonding process. A ewe may steal another lamb because of her exposure to its birth fluids. She may then neglect her own lamb.

The bonding of a mother to her neonate is an intriguing biological process and one that has attracted considerable laboratory attention. If an alien newborn is substituted for a female's own young before she has bonded to her own young, she may then bond to the alien young and reject her own (Klopfer et al., 1964). A number of experiments have been performed to elucidate the nature of this olfactory attachment. Typical of a number of studies is one experiment on the ancient Soay sheep, showing that ablation of the olfactory bulbs of the dams before birth caused a marked reduction in licking of lambs at birth, and strange lambs were often accepted and nursed (Baldwin and Shillito, 1974). Elimination of the dam's sense of smell after the young are born, however, tends to make the dam reject all young, even her own.

Additional work suggests that a mother chemically labels her own kid in some way and rejects alien young because the label is not her own. The labeling may occur through a mother licking her kid and transferring her own unique rumen microflora to the kid's body surface. Another possibility is that the mother's milk imparts unique characteristics to the kid's mouth, or to the kid's body, or to the eliminative products of the young through milk digestion (Gubernick et al., 1979; Gubernick, 1980).

There is no denying the importance of olfaction for a mother in recognizing her young. Although the most attention is given to the anal region, mother sheep can recognize young by either the head or tail region (Alexander, 1978). Interestingly, using strong odoriferous substances does not mask lamb odor or prevent recognition (Alexander and Stevens, 1982). An experiment with sheep indicates that removing olfactory cues from lambs does not promote acceptance of alien lambs by ewes. When lambs aged two to eight days of age were thoroughly washed and anesthetized so that ewes could not use visual or auditory cues to recognize them, ewes rejected their own lambs as well as strange ones. When their own lambs were not anesthetized but were scrubbed, the ewes initially responded positively to their own lambs but then temporarily repelled them when the lambs attempted to nurse (Alexander and Stevens, 1981). A logical next step, in light of the labeling experiment on goats, would be to attempt to label an alien scrubbed lamb with a ewe's own saliva or milk to see if she would then accept it.

The natural rejection of strange young is of practical consequence in situations in which a farmer wants to find a foster mother for young. If one anticipates this need in advance of a mother's own birth, the process is fairly easy. One report describes the technique of putting as many as four preselected calves with a cow (for multiple-calf suckling and rearing) after the cow's own calf was removed at birth before she had any contact with it. The calves, ranging up to 10 days in age, were wiped with amniotic fluid and introduced to the cow within two minutes after parturition. A rapid and specific bond was developed between the cow and her foster calves (Hudson, 1977). This procedure was found to give more uniform growth rates for the calves than the usual fostering method of repeatedly placing partially starved calves with cows until the cows eventually accept them.

A different problem arises in flocks of sheep in which there is a high incidence of multiple births and frequent occurrence of orphaned and abandoned lambs, making it desirable to find foster mothers for lambs several days after a ewe has lambed.

One long-established technique used by sheep herders in fostering is to rub feces or birth fluids from the dead lamb onto the lamb to be fostered. Placing the skin of the ewe's own dead lamb over the alien lamb, so-called skin grafting, will facilitate fostering. Sheep farmers stress that it is important to be sure that the anogenital skin and the head skin are included in the orphan's skin covering. Placing a stockinette that has been passed over one of a ewe's own lambs onto an alien lamb has also induced acceptance of alien lambs (Price et al., 1984). It has been found that simply forcing a ewe and a strange lamb together in a pen for several days, and feeding the young by hand until it is accepted by the ewe, may also work (Hersher et al., 1963). There is also some work showing that tranquilizing a ewe while it is presented with a strange lamb will result in the lamb's being accepted, even after the drug has worn off (Neathery, 1971). This procedure works only if the foster ewe has lost her own lamb. If she has one lamb, she will reject the alien lamb once the tranquilizer wears off (Tomlinson et al., 1982).

An entirely different approach to fostering alien lambs gets at the physiological mechanisms involved. It was found that inflating a balloon in the vagina induced acceptance of alien lambs in ewes 2 to 3 hours postpartum (Keverne et al., 1983). This is a modern variant of the technique used by nomadic cattle herdsmen to induce acceptance of alien calves (Lott and Hart, 1983). Apparently, pressure against the cervix activates a neuroendocrine reflex releasing a hormone such as oxytocin, which induces acceptance.

Milk Ejection Reflex

It is well known that when a nursing animal massages the mammary gland or nurses on a teat, a neuroendocrine reflex is evoked that causes the ejection of milk from the mammary alveoli into the large gland sinuses or gland cisterns from which it can be withdrawn. Massaging or washing the udder prior to milking has the same effect. The reflex is initiated by stimulation of the sensory nerve endings of the skin of the teat and results in the release of the hormone oxytocin from the posterior lobe of the pituitary gland.

In dairy cattle this milk letdown reflex is susceptible to classical conditioning such that the sound or sight of a farmer moving milking cans or in other ways preparing for milking frequently causes the milk to be let down. In fact, visual and auditory stimuli that are closely associated with milking or nursing may result in the buildup of pressure in the mammary gland to such an extent that milk drips through the teat orifice.

The milk letdown reflex is not the same in all species, nor is it as easily conditioned in all domestic animals. In zebu breeds of cattle the reflex cannot be activated in the absence of a suckling calf. The milk letdown reflex does not occur in response to human activity, even with the same person milking a cow year after year. Thus, to induce milk letdown in zebu-type cattle kept for nursing, it is necessary to have the calf suckle first. The calf is then removed in order to obtain milk for human consumption (Figure 4-3). A procedure to induce milk ejection when a calf has died is to make a dummy of the dead calf's head and manipulate it in bunting movements against the udder to deceive the cow (Folley, 1969). Another procedure is to blow air into the vagina, which also leads to the release of oxytocin and induces milk letdown (Debackere and Peters, 1960). In the Fulani, a cattle culture, blowing into the vagina of cows is used to induce milk letdown as well as to facilitate adoption of the newborn as mentioned above (Lott and Hart, 1983). Not all domestic species have the same pattern of milk ejection. In lactating goats oxytocin is not always found in the bloodstream, and milk yields are normal irrespective of whether or not oxytocin is secreted (McNeilly, 1972).

Figure 4-3   Milk letdown will not occur in zebu-type cattle herded by the nomadic Fulani unless the calf is allowed to nurse for a short while before milking begins.

Since the milk ejection reflex can be conditioned in dairy cows, it is not surprising that the reflex can also be inhibited by certain environmental stimuli. In terms of milk production in a dairy herd, this can be quite significant. Numerous studies have revealed that emotionally disturbing stimuli will inhibit milk ejection. There is even some evidence that the actual secretion of milk, as well as milk letdown, may be inhibited. Indeed, the "contented cow" does give more milk. One could even argue that playing music in the milking parlor has a beneficial effect on milk production because the music contributes to a consistency in the environment and becomes part of a cluster of stimuli that condition the milk ejection reflex.

A disturbing environment can inhibit milk letdown and lower milk yield. One possible disturbance is adding new cows to a herd. The emotional disturbance lasting until the new cows have found their place in the rank order, may have an effect on milk yield (Brakel and Leis, 1976). The actual quantitative assessments of such effects are limited; some experiments point to a possible reduction of up to 5 percent in milk production when cows are exchanged between groups, but other studies reveal no effect. One would expect that the new cows, and those with which the new cows interact in competition for rank, would be affected the most (Hart, 1980).

A more direct emotional disturbance affecting the milk ejection reflex is electric shock, which a cow might receive from a milking machine or from metal stanchions. Cows apparently can detect as little as one-half of a volt. The electrical activity can originate from miles away and be quite sporadic, and it is therefore difficult to pinpoint or control. There is evidence that if the shocks are small enough, milk production may be impaired even though a cow's external behavior may indicate no disturbance.

What about the effects of a farmer's personality on milk production? According to some observations, the calm and quiet farmer is likely to obtain higher yields from his cows than the loud extrovert. Having a strange person present can also be inhibiting. It all boils down to the way the cows are handled. If the farmer gets annoyed easily and takes out his financial frustrations on the cows or allows his dog to harass the cows, the cows may react, and milk production is likely to be reduced (Seabrook, 1972).

The Nursing-Suckling Relationship

In dairy cattle and goats it is common to wean the newborn after two to three days of contact with the mother. By this time the newborn has obtained the colostrum, which is the milk secreted just after birth that is high in maternal antibodies. Calves and kids are then raised on a milk substitute. Dairy calves are sometimes removed from cows almost immediately after birth and fed colostrum artificially to eliminate the emotional stress on the mother and young of removing the calf later.

Where milk is supplied artificially in the form of commercial nipples attached to bottles or community buckets, it is delivered much more rapidly to the young than it would be from a real mother. As mentioned previously, young animals have a built-in tendency to nurse irrespective of hunger, and when milk is rapidly obtained with only a short duration of nursing, the young have a tendency to engage in excessive nonnutritive sucking. If communally raised, dairy calves that are bottle-fed tend to nurse on each others ears, genital organs, and other parts of the body (Wood et al., 1967). Since dairy cattle have much larger udders and more free-flowing teats than one could ever find in nature, even calves that are left with their dams are sometimes likely to engage in nonnutritive nursing.

Suckling Episodes   The number of suckling episodes has received some attention. The mean number of nursings per 24 hours reported for Hereford cattle is eight to 12 times, with nursing episodes usually lasting from six to 10 minutes (Ewbank, 1969). Cattle tend to nurse the least frequently of domestic animals and have prolonged episodes. Sheep, horses, and swine tend to nurse hourly in episodes of a minute or less (Ewbank, 1967; Tyler, 1972; Barber et al., 1955).

Anyone who has watched young animals nurse may be aware of the fact that sucking movements are very rhythmical. There are peripheral signs of sucking movements suggesting that a central timing mechanism regulates the rhythmic actions of sucking. Goat kids, for example, wag their tails rhythmically while feeding at a rate that is the same as that of sucking movements (Wolff, 1973).

In herds of beef cattle, nursing calves are sometimes left behind as a "nursery" group while most of the cows in the herd go off to graze or seek water. One cow usually stays behind to tend the nursery. One would think that cows take turns tending calves over sequential days, but this has not been documented.

Followers and Hiders   In beef cattle, sheep, swine, and horses, when young animals are allowed to nurse, they spend most of their time within a very short distance of the mother. In the wild relatives the young derive protection from predators by proximity to their mother and the herd. Generally the terrain is smooth enough to allow the young to follow their mother. As they grow older they gradually spend more time away. It is estimated in the horse, for example, that the foal is within five meters of the mare about 95 percent of the time in the first week of life; at eight months of age, foals are that close only 20 percent of the time (Houpt and Wolski, 1979). Comparable figures probably apply to cattle and sheep.

In contrast to the follower species, there is a different type of mother-infant relationship seen in goats and wild cervid ruminants, such as the deer. For the first few days after parturition, the kid is hidden for as long as eight hours while the mother feeds. At five days of age, the kid stays with the mother as she feeds in a reduced home range. Hiding behavior occurs in those species that inhabit rough terrain where it would be dangerous for the young to follow. The hiding behavior is obviously adaptive, as it reduces the danger inherent in a kid attempting to follow a mother over rough terrain. This behavior has been observed in feral goats in New Zealand. To hide a kid, mothers were observed to lead them to a place under a bush. On returning to the kid, the mother often gave a muted bleat to which the kid responded by rushing out to nurse. A kid might be hidden several times in one day. Toward evening the mother would feed in a restricted area with the kid in sight or alongside (Rudge, 1969). The hiding behavior has been observed in a farm-setting where the interactions between does and their kids were undisturbed. Interestingly, the kids were the ones seen to choose the hiding places (Lickliter, 1984). It is now documented that some ungulate species may be intermediate between the following and hiding classifications, so that either strategy may be used. Goats seem to be in this intermediate group (Lent, 1974; O'Brien, 1984).

Weaning

In wild species, weaning is a gradual process reflecting the increasing frequency of rejection of the young by the mother. Both the mother and the young adjust to the separation over the span of at least a couple of weeks. On farms, by abruptly separating the mother and young we make weaning emotionally traumatic for both the dam and the young. This is evident in the excessive vocalizations of the dams and the frantic attempts of the young to find their mothers. In fact, the young may injure themselves in their attempts to return to their mother. Some attention to this traumatic period may reduce the chance of injury to weanlings and reduce the emotional stress on both mother and young. It may be possible, for example, to allow the dams and offspring to contact each other through open fencing. The barrier holding the weanlings should be secure, and without protruding nails, loose wire, and so forth. Keeping two or more recently weaned animals in the same pen will also reduce the effect of isolation during weaning. This may require delaying the weaning of one animal so as to be able to wean several young at the same time.

Coprophagy in Horses   As the the time of weaning approaches it has been observed that foals will sometimes ingest fecal pellets from their dam or other horses. The function of this behavior is believed to be related to helping the foal establish the intestinal bacterial flora necessary for digestion of forage. However, an interesting series of experiments with rats suggests some additional function. Rat pups that are nearing weaning are very attracted to eating maternal fecal pellets. Through ingestion of the fecal pellets they ingest deoxycholic acid, which protects them against necrotizing enterocolitis, a highly lethal disease of the gastrointestinal tract (Moltz and Kilpatrick, 1978). It is interesting to speculate whether some protective substances might be passed on to young foals by coprophagy, and whether they are particularly attracted to the fecal pellets of their own dams.

■

MATERNAL BEHAVIOR OF POLYTOCOUS SPECIES

For the most part polytocous carnivorous species are not subject to the same kind of predation as the monotocous ungulates. However, they do face a problem in food shortages. They cope with their problems by producing many young, but provide less individual attention for each offspring than is characteristic of monotocous animals.

There is not a strong bond between a mother and individual offspring. This reflects the fact that the animals do not live in herds where the young could otherwise become mixed up. Dogs and cats are born altricial, or immature, which poses no problem since the young do not have to travel with the mother. The fact that the young are born so immature means that mother-young interactions have more pronounced effects on behavioral development. Isolation from the mother soon after birth produces more pronounced effects on dogs and cats than similar isolation in ungulates. Swine, although polytocous, are as precocial as other ungulates at birth.

As with monotocous species, the young of polytocous species gain experience in recognizing and procuring appropriate foods during the lactation period. However, young carnivores must grow and be well developed before they can capture prey. The weaning transition period of dependence on the mother typically continues after lactation ends and lasts until the young can be self-sufficient at hunting (Geyer and Kare, 1981).

Behavior before and during Parturition in Dogs and Cats

In dogs and cats licking of the genital and abdominal areas increases during the later stages of gestation. In both species nest building is quite variable among individuals and usually is not very pronounced. Beds for queening or whelping are often provided by the owners. If left on their own, cats can be expected to engage in more nest building than dogs. In dogs any attempts at nest building usually occur 24 to 48 hours before parturition and, at most, consist of scratching up rugs, towels, rags, or clothing in some corner of the house. In rodents and lagomorphs nest building is quite elaborate. For example, a female rabbit plucks hair from her body a few days before parturition and uses it to line a nest that has been previously constructed of grass or sticks.

The parturient female becomes restless about 24 hours before parturition. In addition to an increased amount of vocalization, this behavior frequently follows the pattern of lying down, standing, walking about, changing position, and lying down again. In addition there are also some physiological changes that may be evident, such as increased respiratory rate, loss of appetite, and increased heart rate.

As in monotocous species, the birth of each newborn proceeds through three stages. There is an interval between deliveries during which the female not only rests but also grooms and cleans the newborn and herself. In the first stage, contraction, uterine contractions begin and there is a great deal of straining. Most polytocous species lie down during this stage although they may frequently get up to change position.

In the second stage, delivery, uterine contractions and contractions of the abdominal muscles become more intense and the fetus moves rather rapidly through the birth canal. As the head or buttocks of the fetus appear at the vulva, the female often breaks the fetal membranes with her teeth and, by tugging on the membranes, may actually pull the fetus through the birth canal. Various positions are assumed during this stage of labor. In the dog and cat the female usually lies on her side and may bend her head to the rear quarters through her back legs (Figure 4-4). Once the newborn has passed through the birth canal the mother rapidly consumes the fetal membranes and begins licking the newborn vigorously. This usually causes the first respiratory

Figure 4-4 Aspects of postparturient maternal behavior in the dog: Chewing away of the placental membrane (top); biting off the umbilical cord (middle); and thoroughly licking the newborn (bottom).

movements of the newborn. It is a common practice for dog breeders to rub down the body of a newborn puppy and stimulate respiration if a bitch is not prompt in this behavior.

In the third stage of labor, delivery of the placenta, the mother continues to lick and groom the newborn. The placenta is readily eaten by the mother. Placentophagy obviously has different functions for carnivores than herbivores. The behavior is necessary housekeeping (nestkeeping) in removing material that would otherwise decay. Placentas also provide some food for mothers, allowing them to stay with the newborn longer. While eating the placenta, the mother usually bites off the umbilical cord. The mother may even appear to become so engrossed in eating the placenta and umbilical cord that she chews into the abdominal wall of a newborn, although this happens fairly infrequently.

As the mother licks and grooms the newborn she concentrates more on the anogenital region. This usually evokes urination and the first defecatory movement. The material that is passed in the first defecation of the infant is meconium and represents secretory material and dead cells of the gut lining.

The interval between deliveries follows the third stage of labor. During this interval the mother not only continues to lick and groom the newborn animals as well as her own genital region but also cleans the bedding that has been soiled with amniotic fluids. Parturition may be a long, drawn-out process in the case of some dogs, with the duration of time between delivery of the first and last fetus lasting as long as 12 to 16 hours. In other animals the birth of the entire litter may take no longer than two hours.

In between deliveries and following delivery of the last fetus, the mother may attempt to retrieve any young that have strayed from the nest. A common form of retrieving in dogs is for the mother to lick the head region of the young pup that has strayed from the group. In response to this tactile stimulation, the newborn moves toward the stimulation and is thus directed back to the rest of the litter (Figure 4-4). Most of the newborn begin to nurse soon after they have been delivered, even before delivery of the last fetus.

Postparturient Behavior of Dogs and Cats

Although it may not be apparent to us, newborn puppies and kittens enter an environment that is quite highly structured. They cannot see, hear, or move about very well, but their olfactory and thermal receptive systems are well developed (Rosenblatt, 1976). Research on kittens has demonstrated that there are both thermal and olfactory gradients between the home or nest and other areas into which a kitten might wander. The thermal gradient is maintained by the mother, who spends most of her time in the nest. This gradient is continuous from her abdomen, out her legs, and onto the floor of the nest area. Kittens can sense and follow this thermal gradient to the mother (Freeman and Rosenblatt, 1978a).

There is also an olfactory gradient that becomes more diluted as kittens get farther from the mother. By the end of the first week after birth almost all kittens can find their way to the nest by olfaction. When they do get home, the warmth of the nest causes them to settle down (Freeman and Rosenblatt, 1978a). Further work shows that kittens seem to recognize and orient toward two nest odors: one is general to all mothers, and one is specific to the kitten's own mother. Given a choice, they will orient to the smell of their own mother and not to that of a strange mother (Freeman and Rosenblatt, 1978b).

For the first three weeks of life mother dogs and cats continue to lick and groom each newborn. Much of this grooming is directed toward the anogenital region, which evokes urination and defecation on the part of the young. Urine and fecal material are consumed by the mother, in this way the bed is kept clean. The behavior of licking the anogenital area and consuming the fecal material of the young is thought by some canine authorities to predispose dogs to coprophagy as adults, especially when they are confined.

Whether a mother licks all the young in her litter equally is not known. There may be a difference in the attention given males and females. Rat mothers are reported to lick the anogenital region of males considerably more than that of females (Moore and Morelli, 1979). This effect has not been examined in dogs or cats.

As the young are able to move about, the anogenital licking subsides, and the young deposit feces and urine on one side of the bed or in a part of the room away from the bed.

Much of postparturient behavior involves nursing the young. The importance of the newborn in evoking the neuroendocrine reflex involved in milk ejection was discussed when we considered monotocous species. Ejection of milk must take place before there is milk available to the suckled young. The vigorous sucking movements and the treading and kneading of the mammary glands as seen especially in kittens, brings about the milk ejection. Interestingly, massage of the mammary gland and the suckling action on just one or a few teats brings about milk ejection of the entire set of mammary glands (Drewett and Trew, 1978). Hence, we see the reason why puppies and kittens may nurse on two or three nipples in succession.

Progression of the Nursing-Suckling Relationship   This aspect of mother-young interactions occurs in three phases. In the first phase the mother plays a major role in initiating nursing by hovering over the litter or lying near the group of newborn and arousing them to nursing activity by licking them. In response to this licking, the young nuzzle into the fur of the female, find a teat, and nurse. As time progresses the newborn become very adept at finding teats and responding to the mother's solicitous behavior. A mother will leave the bed at periodic intervals to feed, stretch, eliminate, and, in the case of household pets, to go for a run or car ride. Dogs seem somewhat more willing to leave the bed for prolonged periods than cats. The first phase of lactation lasts until the pups or kittens are about two and a half weeks of age, which is the time when the eyes and ears of the newborn are open and they are able to move about more easily.

In the second phase the young are able to leave the bed and to interact with the mother outside of the bed. In this phase the young initiate most nursing episodes; these may take place inside the bed or outside. The mother usually cooperates by resting with her nipples exposed for varying periods of time.

In the third phase of the nursing-suckling relationship the young are able to take food from other sources. During this phase practically all of the nursing is initiated by the young, and as time progresses, the mother frequently evades nursing attempts by the young. For example, the mother may lie with her mammary region against the floor, or she may climb to places where the young cannot reach her. Weaning is brought about toward the end of this phase by the mother becoming less available to the young (Schneirla et al., 1963). The progression of the nursing-suckling interactions through the three stages can be checked or even reversed by taking away some of the older young and substituting young that are more immature (Korda and Brewinska, 1977a, 1977b).

Nipple Location and Control of Suckling   Although the location of a nipple by the newborn of livestock species appears to be mainly through tactile cues, there is reason to believe that newborn dogs and cats are dependent on olfactory cues for nipple location. Saliva appears to play an important role in this process, because often a mother arouses young by licking their faces. As the young start to move the mother will lick her own nipples as well as the newborn. She thus lays down a saliva scent trail that they follow by lateral head-scanning movements until they contact a nipple (Blass and Teicher, 1980).

The importance of saliva as an olfactory cue for the blind newborn to use in locating a nipple has been studied in detail in rats. Chemical washing of the nipple virtually eliminates suckling by neonatal rat pups. Normal attachment to the washed nipples can be induced by painting the nipples with a distillate of the nipple wash, parturient-mother saliva, or amniotic fluids (Teicher and Blass, 1976, 1977).

These experiments are interpreted as evidence that chemical cues are vital in leading rat pups to a nipple. It is not surprising that both maternal saliva and amniotic fluid are effective in enabling the newborn to find a nipple since the mother's tongue, in addition to carrying saliva, will also carry amniotic fluid from licking the newborn or her own anogenital area. Interestingly, saliva and amniotic fluids not only lead a rat pup to the nipples but also elicit mouth opening and tongue extensions. After the first couple of nursings, saliva from the newborn left on the nipple from previous nursings will do the same thing. There is evidence that secretion of a material from the nipple, induced by oxytocin production, also facilitates a rat pup's ability to find and attach to a nipple (Singh and Hofer, 1978). Eventually, of course, tactile qualities of the nipple alone evoke sucking.

Many pet owners have noted that sticking a newborn kitten's or puppy's mouth onto a nipple will not evoke attachment and sucking no matter how forcefully this is done. Why? If the mother has not licked her nipple area, or only licked some nipples, the uncoated nipples will not have the olfactory stimulus needed to evoke suckling behavior. The logical thing to do is to wait for the mother to lick her nipple area or at least first rub mother saliva and/or amniotic fluid on the nipple and onto the newborn's muzzle before placing the newborn on the nipple.

Despite the fine-tuned sensory controls governing nipple attachment and the initiation of suckling, once nursing activity starts, it is pretty much free-running. Nursing activity does not depend on the amount of milk in the stomach. Even rat pups given milk by stomach tube to the point where there is observable stomach distention will rapidly attach to a nipple and suck. In older rat pups milk deprivation and stomach filling do affect suckling attempts, but in newborn animals suckling behavior occurs on its own independent of milk deprivation or satiation (Rosenblatt, 1976). An experiment with cats also substantiates this concept. Kittens that were fed by stomach tube were allowed contact for two hours a day with female cats that were either lactating or nonlactating. The dry-suckling group spent as much time suckling as the milk-suckling group until they were about three weeks old (Koepke and Pribram, 1971).

There appears to be considerable variation between species, and among individual litters within a species, as to whether

young form attachments to certain nipples. In pigs the ownership of a nipple is quite the rule. In cats the literature emphasizes
that kittens tend to prefer certain teats (Ewer, 1961; Schneirla
et al., 1963). However, cat breeders have observed many litters
in which kittens are indiscriminate in the nipples they choose
(Hart, 1972; Beaver, 1980). In dogs one rarely finds attachment
of pups to specific nipples (Rheingold, 1963).

Communication between Mother and Neonates   Our casual observations of interactions between mother and young probably
do not reveal the complexity of communication that is occurring. The discovery of a chemical substance that guides young
rat pups to the nipple is an example. Another example is the
maternal pheromone produced by the fecal bacteria of the rat
mother at about the time when her pups start to venture from
the nest. The pheromone has the effect of attracting the pups
to the nest (Leon, 1974, 1979) and of inducing them to consume
a protective substance in fecal pellets (Moltz and Kilpatrick,
1978).

  In another area, research is beginning to reveal much about
ultrasonic communication in rodents (Geyer, 1979). Rodent
pups in distress are known to produce two different types of calls.
One is a response to cold, which has the effect of attracting the
mother to the pups and stimulating searching and retrieving
behavior. The other sound is emitted when pups are handled
or subjected to aversive tactile stimulation; these sounds cause
a mother to withdraw. In nature such sounds are produced when
a pup is being sat on or smothered by the mother (Brown, 1976;
Noirot, 1972). Ultrasonic cries from infant rats may also stimulate the release of prolactin in mothers (Terkel et al., 1979).
The discovery of two different kinds of ultrasonic auditory messages, which cause approach or withdrawal, suggests a more
complex communication system among young and mother than
we might have anticipated.

  The young of all species of domestic animals emit sounds
when separated from their mother or nest. These sounds are
usually referred to as distress calls. No ultrasonic emissions have
been identified in common domestic animals. Sonographic analyses of repeated calls by kittens reveal that there are definite
individual differences that a mother may recognize but we may
not (Brown et al., 1978). These kitten calls are in general intense
and tonal. A mother cat also emits calls when a kitten is taken

from her nest beyond her reach. This is a low-intensity, gargly vocalization. The call may function as a command for kittens to come to her or merely as a guiding stimulus for kittens that have strayed too far (Brown et al., 1978).

Retrieving   This behavior, which is characteristically associated with maternal behavior in cats, apparently is at its peak about one week after parturition. Mother cats typically do not retrieve their young on sight, but rather respond to the young mostly when they vocalize. Very often retrieving occurs only when the sounds reach a rather high intensity. Thus, the kittens that are marooned several feet from the nest and emit stress vocalizations are the most likely to be retrieved. Mother cats sometimes shift a litter from one spot to another in response to environmental disturbances. This tendency is strongest between 25 to 35 days after birth (Schneirla et al., 1963).

When being carried, young kittens become immobile and assume a posture with the tail drawn up and the legs raised (Figure 4-5). By remaining immobile rather than struggling the young facilitate the transportation. Also, with the limbs and tail tucked up close the young makes a compact bundle. This posture is induced by stimuli the young receives when it is grasped by the skin of the neck. Even in adult cats the posture and immobility characteristic of the neonate being carried can be induced by holding a cat by the skin of the neck (Figure 4-5). This procedure has proved to be an effective way of restraining cats for giving injections or taking rectal temperature (Hart, 1975).

Mother dogs are much less easily induced to carry their pups or to relocate a litter. When they do carry a pup, it is by grasping any part of the body such as a front leg or the head and dragging it.

Cross-Fostering   Polytocous species will often readily accept alien young, even of a different age than their own or of a different species. Cross-fostering is taken advantage of at times to save an orphaned young, if a mother with just a few young of similar age can be found. Attempts at cross-fostering orphans may require close attention and possibly a bit of help in adapting the foster animal to suckling from the new mother. In one experiment bearing on problems with fostering, it was found that kittens that were removed from their mother for several days

Figure 4-5 The same immobility reflex assumed by kittens when carried by their mothers can be induced in an adult cat and used as a means of restraint for injections. (From Hart, 1975. Reproduced with permission of Veterinary Practice Publishing Co.)

took a considerable period to resume normal suckling behavior on being returned to the litter. This delay did not represent a rejection by the mother but rather a deficiency on the part of the neonate to orient to the mother (Schneirla et al., 1963).

Transferring Weanlings to Solid Food   In wild carnivores there is a gradual transition from nursing to solid food. Different species have various species-specific ways of facilitating this transition in the wild. Members of a wolf pack, including the dam and sire of the pups, introduce their young to solid food. They bring partially processed food to the pups by regurgitating a freshly consumed meal in the presence of the pups. Wolf pups eagerly "beg" for the food by pawing and biting at the lips of an adult who has just returned to the nest. Later, the pups are introduced to freshly killed small game or parts of a carcass of larger prey. Pups are totally weaned when they can exist on this type of solid food. Later they can engage in hunting with older members of the pack.

As household pets, dogs are provisioned at the time of weaning by their human owners. Nowadays we pay particular attention to providing palatable and easily chewed food to weanling puppies. Disgorging by the bitch is seen only infrequently. In fact, the behavior has probably been selected against because it is somewhat objectionable to people.

The ancestors of cats lived a solitary existence, and it was essential for the young to acquire hunting skills before leaving the nest. Domestic mother cats, if given the opportunity, will often teach their kittens to hunt by initially bringing them killed or injured prey. Later the kittens may go with their mother on a hunting trip. If kittens do not receive this early experience, they are less likely to develop predatory tendencies as adults, but some cats will still hunt, apparently even without the maternal introduction to hunting.

For people who live in rural areas and want to enhance the tendency of their cats to hunt rodents, selecting kittens from litters in which the mother is known to hunt, and being sure that the kittens are left with the mother long enough to learn to hunt, is the best way of ensuring that kittens will become good hunters. Since many cats will hunt even though not hungry, it is often a good idea to feed the cats to keep them around the home.

People who live in a city or suburb may wish to acquire cats that are the least likely to prey on songbirds or on small mammals such as squirrels or chipmunks. They should select kittens from litters in which the mothers are known to be nonhunters. If this is not possible, then kittens should be removed from the litter before they are old enough to learn how to hunt from the mother (Hart, 1976).

Postparturient Behavior of Swine

What hits most behaviorists who have studied mother-young interactions in pigs is the rapidity and ease with which newborn piglets scamper about. Within a minute or two after birth piglets are capable of moving about on their own and freeing themselves of placental membranes. Thus the sow does not rip the umbilical cord with her teeth nor does she lick and groom the newborn. The most prominent aspect of mother-young interactions of swine at the time of parturition is an exchange of vocalizations.

Piglets are able to locate nipples quite easily and begin nursing almost immediately after freeing themselves of placenta membranes and the umbilical cord. Soon after birth they attach to a specific teat. The most anterior teats appear to be favored the most, and there may be fighting by piglets over favored teats in the first few hours of life. The "teat order" often serves as the basis for a later social order among the litter. Sight of the teat, olfaction, and recognition of neighbors are involved in the identification of teats. A piglet may recognize its own smell on a teat, which is left over from rubbing its nose around the teat. The role of neighbor recognition is seen when a row of piglets may move one teat to the right or left, usually displacing a piglet in the process.

Observations of the interaction among piglets during suckling reveal that unethical practices such as poaching on another's teat when the owner is away, or occupying two teats, do occur. If a piglet loses its favorite teat, it is claimed that it may withdraw for days and refuse to suckle from any other teat (McBride, 1963).

In watching piglets nurse, it becomes apparent that each nursing bout has several distinct phases. First they must gather about the sow, either on their own accord or in response to the sow's grunts. They then attach to their preferred teats, after some jostling for position. As they settle down the piglets massage the udder with vigorous, rhythmic, up-and-down movements of their snouts; this type of nursing lasts about a minute. Their vigorous massaging behavior then gives way to slow suckling movements in which the tongue is usually visibly wrapped around the teat. This phase lasts about 20 seconds. Up to this point no milk has been released. Next one can see that the slow suckling movements are replaced by rapid mouth movements as milk starts to flow. During this phase, which lasts for only 10 to 20 seconds, the piglets pull back slightly, their ears are flattened, and their heads seem to move in coordination with the sucking movements. After the 10 to 20 seconds of milk flow, the piglets may resume the slow mouth movements or dart from teat to teat as if to look for more milk. The nursing episode ends with the piglets leaving the udder or falling asleep in position. The sow usually emits rhythmic grunts throughout the nursing episode. The frequency of grunts increases gradually but stops once the pigs stop sucking (Fraser, 1980).

One piglet alone cannot cause milk ejection. What is the adaptiveness of such a high threshold for ejection and the long time lag from when the piglets start working on the udder to the onset of milk flow? The current explanation is that the time lag from udder massage to milk ejection ensures that most of the piglets are in position and already sucking by the time milk flows. The long preliminary phase allows stragglers to reach the udder before milk flow begins; furthermore, the sow's grunts advertise to any absent piglets that a nursing is in progress. These safeguards for ensuring that an individual piglet cannot elicit milk flow, and the time lag to make sure that littermates are prepared, are important to a polytocous species with precocial young that may wander off a bit and miss a nursing. One piglet cannot corner all the milk for itself (Fraser, 1980).

Sows may kill piglets in the first few days after birth by lying on them and smothering them. Farrowing stalls are used to prevent injury of piglets, but even with the stalls the smothering of piglets is still a disheartening and economic problem. Some sows seem to be naturally much more careful in lying down than others. It is quite possible that farmers could selectively breed for sows that are sensitive to the presence of piglets and that respond to the squealing of piglets being hurt. Another problem is that even with well-behaved sows, farmers cannot leave well enough alone. In order to increase the efficiency of breeding stock, they are weaning pigs at earlier ages. Instead of weaning at eight weeks, the old norm, piglets may now be weaned at six or even three weeks of age. The practice, though financially necessary, results in unusually high levels of general activity, aggressiveness, nonnutritional suckling, and faulty regulation of food intake in piglets (Fraser, 1978).

What about putting alien piglets with a sow and her own litter? Because pigs are a species that has precocial young, one might expect a sow to reject alien young as would a ewe or a cow, but because pigs are also a species that gives birth to a litter, one might expect more the open-arms policy of dogs and cats. Not surprisingly what is observed is behavior somewhat in between. Foster piglets are the brunt of some rejection by sows, but they escape being attacked. A sow does apparently identify her young by means of olfaction, and as a sow's piglets get older, alien young are more vigorously rejected. Making sows anosmic by performing olfactory bulbectomy prior to breeding eliminates rejection of alien piglets (Meese and Baldwin, 1975).

An added complication in studying the welfare of alien piglets put with foster mothers is their interaction with resident piglets. The fostering process results in the disruption of sucking efficiency by the entire litter by disturbing the teat order and promoting aggression. The weight gained by the foster piglets is often less than that of nonfoster piglets (Horrell, 1980).

Aggression toward the Newborn and Cannibalism

Mothers may sometimes attack and kill their newborn offspring in the first few days after birth. If they eat part or all of the dead newborn, we refer to the behavior as cannibalism. Killing and cannibalism of the newborn have been observed in all polytocous species and in some primates. We tend to be shocked by the fact that a mother could eat her own young; to kill when necessary might be understandable, but to eat one's offspring seems entirely bizarre. In nature, cannibalism of dead young is not so out of line when one considers that a mother must keep the nest clean of dead young just as she does by eating placentas and consuming eliminated urine and feces of the young. Assuming she does not kill all of her own young, the protein she derives from the sacrificed ones helps the long-term survival of the others.

Can infant killing ever be a normal aspect of the reproductive process? Apparently yes. Postparturient female hamsters, for example, almost always cannibalize a few of their offspring. Cannibalism is a normal part of maternal behavior used to adjust litter size in accordance with the environmental conditions and food supply prevailing at the time of parturition (Day and Galef, 1977).

There is evidence that mothers of other species also adjust their litter sizes such that all surviving young can be adequately cared for. A female mouse has 10 nipples and 10 or fewer pups should receive adequate nutrition. When excessive foster pups are added to the litter, some of the young are killed. The more foster pups added, the more young are killed. What is more, the smaller and presumably weaker pups are the ones weeded out (Gandelman and Simon, 1978).

In other species there is some evidence that cannibalism is brought on by the nutritional needs of the mother. One study of squirrel monkeys found that abortion and cannibalism occurred in females on a low-protein diet but was never observed in females on high-protein diets (Manocha, 1976).

Can any of this information on rodents or squirrel monkeys be of value in understanding cannibalism in dogs, cats, and pigs? The answer is clearly yes. When cannibalism does occur in dogs and cats, it is often the sickly infants that are killed. To rid the nest of sickly infants may protect the remaining litter. Perhaps we are seeing the process of nature at work. Also, there is no particular reason to assume that mother dogs and cats can adequately raise as many young as there are nipples; the cutoff point may be considerably lower.

When cannibalism of normal, healthy young in a normal-size or small litter occurs, many experts feel that it is probably related to the immaturity of the mother, lack of maternal experiences, illness of the newborn, hyperemotionality, or environmental disturbances. There may also be hormonal factors involved that incite a mother to attack her infants. It is known that the placenta produces appreciable amounts of progesterone during pregnancy. With the detachment and expulsion of the placenta at parturition, there is a marked fall in progesterone in the mother's body. Since progesterone has physiological calming properties, the fall in the level of progesterone may precipitate irritability and aggression toward the young, especially if environmental disturbances are occurring at the same time. This is one explanation of postparturient aggression and depression in some women.

Perhaps the best example of the role of emotional disturbance in the killing or cannibalizing the young is in pigs. Sows that are particularly nervous or hyperemotional are recognized by farmers as the most likely to attack and eat piglets soon after birth. If the sow is moderately nervous, a farmer can often calm her and get her to lie down by rubbing the udder area. For more extreme nervousness, a single injection of a tranquilizer often attenuates the nervousness of the sow and leads to milk letdown, which may have been inhibited. Once the piglets are suckling, the sow loses her tendency to attack them, even after the tranquilizer has worn off (Callear and van Gestel, 1971).

A type of normal infanticide, which sometimes involves cannibalism, is found in some wild felids and a few primate species.

Males that have taken over territory from another male will go to a litter of newborn animals and indiscriminately kill them. With the newborn out of the picture, the female will come into estrus again soon, and the male can then sire the next round of offspring. In anecdotes a similar type of infanticide has been reported to occur in domestic tomcats. How much this behavior occurred in the ancestor of the cat is not known, but it certainly does no harm for the owners of a lactating female cat to watch over the litter of kittens if tomcats have access to the kittens.

Pseudopregnancy

Various ramifications of pseudopregnancy are found in most polytocous species. In the domestic cat and laboratory rodents, pseudopregnancy results from a sterile mating and is believed to be caused by prolonged secretion of progesterone by the ovary. The commonest indication of pseudopregnancy is an enlargement of the abdomen and some development of the mammary glands.

Dogs are unique among domestic species in going through pseudopregnancy regardless of whether or not they have experienced a sterile mating. Also, dogs are the only animals, other than human females, that occasionally display an elaborate degree of maternal behavior as a manifestation of pseudopregnancy. In dogs the syndrome starts with the bitch showing enlargement of the mammary area and abdomen. The mammary glands may develop to the point of secreting milk. Not knowing whether their dog was mated or not, many dog owners fully expect that the bitch is pregnant. In most cases pseudopregnancy subsides a couple of weeks before, or at the time of, expected parturition. In some bitches the phenomenon continues past the date of expected parturition, with the female displaying behavioral signs of parturition, including abdominal contractions and straining. Soon afterward the dog may collect one or two stuffed toys and treat them as newborn puppies. She may lick and hover over the toys as if to nurse them (Figure 4-6). This type of maternal behavior may continue for almost as long as a normal lactation. When the expected time of weaning arrives, the bitch then abandons the adopted toys, and her behavior returns to normal.

The behavioral aspects of canine pseudopregnancy appear to be highly functional in a wolf pack, in which usually only the

Figure 4-6 The pseudopregnancy syndrome in dogs can be accompanied by the adoption of a stuffed toy with attempts to nurse it (top) and defense of the toy against intruders (bottom).

dominant female gives birth. Other females of the pack, which are usually genetically related aunts of the pups, display maternal behavior as part of pseudopregnancy and can serve as nursemaids. In this way they contribute to the survival of offspring related to them and enhance, indirectly, their own reproductive output (Voith, 1980).

We have seen how different aspects of maternal behavior are closely tied into the biological and ecological needs of a species' life style. Tight bonding or loose bonding between mother and young relate to the species-typical social environments. Communication between mothers and infants can utilize channels that are not monitored by our own senses. Hence, this area of domestic animal behavior remains unexplored. Observations of rodents and wild animals provide clarification that the antithesis of maternal care—infanticide—is sometimes adaptive. In the domestic dog we see an example of elaborate maternal responses, representing pseudopregnancy, being displayed by females that have no young.

Despite the differences among species in maternal behavior, in all instances the interactions between mother and young, and among the young, provide experiences that contribute to an animal's permanent behavioral repertoire. Some of the more interesting and important aspects of such early experiences are the focus of Chapter 7.

References

ALEXANDER, G. 1977. Role of auditory and visual cues in mutual recognition between ewes and lambs in Merino sheep. *Appl. Anim. Ethol.* 3:65–81.

ALEXANDER, G. 1978. Odour, and the recognition of lambs by Merino ewes. *Appl. Anim. Ethol.* 4:153–158.

ALEXANDER, G., AND E. E. SHILLITO. 1977. Importance of visual clues from various body regions in maternal recognition of the young in Merino sheep (*Ovis aries*). *Appl. Anim. Ethol.* 3:137–143.

ALEXANDER, G., AND E.E. SHILLITO-WALSER. 1978 Visual discrimination between ewes by lambs. *Appl. Anim. Ethol.* 4:81–85.

ALEXANDER, G., AND D. STEVENS. 1981. Recognition of washed lambs by Merino ewes. *Appl. Anim. Ethol.* 7:77–86.

ALEXANDER, G., AND D. STEVENS. 1982. Failure to mask lamb odour with odoriferous substances. *Appl. Anim. Ethol.* 8:253–260.

ARNOLD, G. W., C. A. P. BOUNDY, P. D. MORGAN, AND G. BARTLE. 1975. The roles of sight and hearing in the lamb in the location and discrimination between ewes. *Appl. Anim. Ethol.* 1:167–176.

ARNOLD, G. W., AND P. J. PAHL. 1974. Some aspects of social behaviour in sheep. *Anim. Behav.* 22:594–600.

ARNOLD, G. W., S. R. WALLACE, AND R. A. MALLER. 1979. Some factors involved in natural weaning processes in sheep. *Appl. Anim. Ethol.* 5:43–50.

BALDWIN, B., AND E. SHILLITO. 1974. The effects of ablation of the olfactory bulbs on parturition and maternal behaviour in Soay sheep. *Anim. Behav.* 22:220–223.

BARBER, R. S., R. BRAUDE, AND K. G. MITCHELL. 1955. Studies on milk production of large white pigs. *J. Agric. Sci.* Camb. 46:97–118.

BEAVER, B. 1980. *Veterinary aspects of feline behavior.* St. Louis: Mosby.

BLASS, E. M., AND M. H. TEICHER. 1980. Suckling. *Science* 210:15–21.

BRAKEL, W. J., AND R. A. LEIS. 1976. Impact of social disorganization on behavior, milk yield, and body weight in dairy cows. *J. Dairy Sci.* 59:716–721.

BROWN, A. M. 1976. Ultrasound and communication in rodents. *Comp. Biochem. Physiol.* 53A:313–317.

BROWN, K. A., J. S. BUCHWALD, J. R. JOHNSON, AND D. J. MIKOLICH. 1978. Vocalization in the cat and kitten. *Develop. Psychobiol.* 11:559–570.

CALLEAR, J. F. F., AND J. F. E. VAN GESTEL. 1971. An analysis of the results of field experiments in pigs in the UK and Ireland with the sedative neuroleptic azaperone. *Vet. Rec.* 89:453–458.

DAY, C., AND B. GALEF. 1977. Pup cannibalism: One aspect of maternal behavior in golden hamsters. *J. Comp. Physiol. Psychol.* 91:1179–1189.

DEBACKERE, M., AND G. PETERS. 1960. The influence on vaginal distention in milk ejection diuresis in the lactating cow. *Arch. Int. Pharmacodyn.* 123:462–471.

DREWETT, R. F., AND A. M. TREW. 1978. Milk ejection of the rat as a stimulus and a response to the litter. *Anim. Behav.* 26:982–987.

ESTES, R. D. 1967. The comparative behavior of Grant's and Thomson's gazelles. *J. Mammal.* 48:189–209.

ESTES, R. D., AND R. K. ESTES. 1979. Birth and survival of wildebeest calves. *Z. Tierpsychol.* 50:45–95.

EWBANK, R. 1967. Nursing and suckling behaviour amongst Clun Forest ewes and lambs. *Anim. Behav.* 15:251–258.

EWBANK, R. 1969. The frequency and duration of the nursing periods in single-suckled Hereford beef cows. *British Vet. J.* 125:1–2.

EWER, F. R. 1961. Further observations on suckling behaviour in kittens together with some general considerations of the interrelations of innate and acquired responses. *Behaviour* 18:247–160.

FOLLEY, S. 1969. The milk-ejection reflex. A neuroendocrine theme in biology, myth and art. *Proc. Soc. Endocrinol.* 44:x–xx.

FRASER, D. 1978. Observations of the behavioural development of suckling and early-weaned piglets during the first six weeks after birth. *Anim. Behav.* 26:22–30.

FRASER, D. 1980. A review of the behavioural mechanism of milk ejection of the domestic pig. *Appl. Anim. Ethol.* 6:247–255.

FREEMAN, N. C. G., AND J. S. ROSENBLATT. 1978a. The interrelationship between thermal and olfactory stimulation in the development of home orientation in newborn kittens. *Devel. Psychobiol.* 11:437–457.

FREEMAN, N. C. G., AND J. S. ROSENBLATT. 1978b. Specificity of litter odors in the control of home orientation among kittens. *Devel. Psychobiol.* 11:459–468.

GANDELMAN, R., AND N. G. SIMON. 1978. Spontaneous pup-killing by mice in response to large litters. *Devel. Psychobiol.* 11:235–241.

GEYER, L. A. 1979. Olfactory and thermal influences on ultrasonic vocalization during development in rodents. *Amer. Zool.* 19:420–431.

GEYER, L. A., AND M. R. KARE. 1981. Taste and food selection in the weaning of nonprimate mammals. In *Food as an environmental factor in the genesis of human variability*, pp. 69–82, eds. D. Walcher and N. Kretchmer. New York: Masson Publishing Co.

GUBERNICK, D.J. 1980. Maternal imprinting or maternal labeling in goats. *Anim. Behav.* 28:124–129.

GUBERNICK, D. J., K. C. JONES, AND P. H. KLOPFER. 1979. Maternal imprinting in goats. *Anim. Behav.* 27:314–315.

HART, B. L. 1972. Maternal behavior. II. The nursing-suckling relationship and the effects of maternal deprivation. *Feline Pract.* 2(5):6–10.

HART, B. L. 1975. Handling and restraint of the cat. *Feline Pract.* 5(2):10–11.

HART, B. L. 1976. Behavioral aspects of raising kittens. *Feline Pract.* 6(6):8,10,20.

HART, B. L. 1980. Behavior and milk production. *Bovine Pract.* 1(2):8–10.

HERSHER, L., J. B. RICHMOND, AND A. U. MOORE. 1963. Maternal behavior in sheep and goats. In *Maternal behavior in mammals*, pp. 203–232, ed. H. L. Rheingold. New York: Wiley.

HORRELL, R. I. 1980. Cross-fostering in piglets. *Appl. Anim. Ethol.* 6:303–304.

HOUPT, K. A., AND T. R. WOLSKI. 1979. Equine maternal behavior and its aberrations. *Equine Pract.* 1(1):7–20.

HUDSON, S. J. 1977. Multiple fostering of calves onto nurse cows at birth. *Appl. Anim. Ethol.* 3:57–63.

HUDSON, S. J., AND M. M. MULLORD. 1977. Investigations of maternal bonding in dairy cattle. *Appl. Anim. Ethol.* 3:271–276.

KEVERNE, E. B., F. LEVY, P. POINDRON, AND D. R. LINDSAY. 1983. Vaginal stimulation: An important determinant of maternal bonding in sheep. *Science* 219:81–83.

KLOPFER, P. H. 1971. Mother love: What turns it on? *Am. Sci.* 59:404–407.

KLOPFER, P. H., D. K. ADAMS, AND M. S. KLOPFER. 1964. Maternal "imprinting" in goats. *Proc. Natl. Acad. Sci.* 52:911–914.

KLOPFER, P. H., AND M. KLOPFER. 1977. Compensatory responses of goat mothers to their impaired young. *Anim. Behav.* 25:286–291.

KOEPKE, J., AND K. PRIBRAM. 1977. Effect of milk on the maintenance of sucking behavior in kittens from birth to six months. *J. Comp. Physiol. Psychol.* 75:363–377.

KORDA, P., AND J. BREWINSKA. 1977*a*. Effect of stimuli emitted by sucklings on tactile contact of bitches with sucklings and on number of licking acts. *Acta Neurobiol. Exp.* 37:99–116.

KORDA, P., AND J. BREWINSKA. 1977*b*. The effect of stimuli emitted by sucklings on the course of their feeding by bitches. *Acta Neurobiol. Exp.* 37:117–130.

KRISTAL, M. B. 1980. Placentophagia: A biobehavioral enigma. *Neurosci. Biobehav. Rev.* 4:141–150.

KRUUK, H. 1972. *The spotted hyena.* Chicago: University of Chicago.

LEIGH, H., AND M. HOFER. 1973. Behavioral and physiologic effects of littermate removal on the remaining single pup and mother during the preweaning period in rats. *Psychosom. Med.* 35:497–508.

LENHARDT, M. L. 1977. Vocal contour cues in maternal recognition of goat kids. *Appl. Anim. Ethol.* 3:211–219.

LENT, P. C. 1974. Mother-infant relationships in ungulates. In *The behaviour of ungulates and its relationship to management*, pp. 14–55, eds. V. Geist and F. Walther. Morges, Switzerland: IUCN.

LEON, M. 1974. Maternal pheromone. *Physiol. Behav.* 13:441–453.

LEON, M. 1979. Mother–young reunions. *Prog. Psychob. Physiol. Psychol.* 8:301–334.

LICKLITER, R. E. 1982. Effects of a post-partum separation on maternal responsiveness in primiparous and multiparous domestic goats. *Appl. Anim. Ethol.* 8:537–542.

LICKLITER, R. E. 1984. Hiding behavior in domestic goat kids. *Appl. Anim. Behav. Sci.* 12:245–251.

LOTT, D. F. 1973. Parental behavior. In *Perspectives on animal behavior*, ed. G. Bermant. Glenview, Ill.: Scott, Foresman.

LOTT, D. F., AND B. L. HART. 1973. The Fulani and their cattle: Applied behavioral technology in a nomadic cattle culture and its psychological consequences. *National Geographic Society Research Reports* 14:425–430.

MANOCHA, S. 1976. Abortion and cannibalism in squirrel monkeys (*Saimiri sciureus*) associated with experimental protein deficiency during gestation. *Lab. Anim. Sci.* 26:649–650.

MC BRIDE, G. 1963. The "teat-order" and communication in young pigs. *Anim. Behav.* 11:53–56.

MC NEILLY, A. 1972. The blood levels of oxytocin during suckling and hand milking in the goat with some observations on the pattern of hormone release. *J. Endocrinol.* 52:177–188.

MEESE, G. B., AND B. A. BALDWIN. 1975. Effects of olfactory bulb ablation on maternal behaviour in sows. *Appl. Anim. Ethol.* 1:379–386.

MOLTZ, H., AND KILPATRICK, S. J. 1978. Response to the maternal pheromone in the rat as protection against necrotizing enterocolitis. *Neurosci. Biobehav. Res.* 2:277–280.

MOORE, C. L., AND G. A. MORELLI. 1979. Mother rats interact differently with male and female offspring. *J. Comp. Physiol. Psychol.* 93:677–684.

MORGAN, P. D., C. A. P. BAUNDY, G. W. ARNOLD, AND D. R. LINDSAY. 1975. The roles played by the senses of the ewe in the location and recognition of lambs. *Appl. Anim. Ethol.* 1:135–150.

NEATHERY, M. W. 1971. Acceptance of orphan lambs by tranquilized ewes. *Anim. Behav.* 19:75–79.

NOIROT, E. 1972. Ultrasounds and maternal behavior in small rodents. *Develop. Psychobiol.* 5:317–387.

O'BRIEN, P. H. 1984. Leavers and stayers: Maternal postpartum strategies. *Appl. Anim. Behav. Sci.* 12:233–243.

PRICE, E. O., G. C. DUNN, J. A. TALBOT, AND M. R. DALLY. 1984. Fostering lambs by odor transfer: The substitution experiment. *J. Anim. Sci.* 59:301–307.

RHEINGOLD, H. L. 1963. Maternal behavior in the dog. In *Maternal behavior in mammals*, pp. 169–202, ed. H. L. Rheingold, New York: Wiley.

ROSENBLATT, J. A. 1967. Nonhormonal basis of maternal behavior in the rat. *Science* 156:1512–1514.

ROSENBLATT, J. S. 1976. Suckling and home orientation in the kitten: A comparative developmental study. In *Biopsychology of development*, pp. 345–410, eds. F. Tobach, L. R. Aronson, and E. Shaw. New York: Academic.

RUDGE, M. 1969. Mother and kid behaviour in feral goats (*Capra hircus* L.). *Z. Tierpsychol.* 27:687–692.

SCHALLER, G. B. 1972. *The Serengeti lion.* Chicago: University of Chicago.

SCHILLIOT, E. E. 1975. A comparison of the role of vision and hearing in lambs finding their own dams. *Appl. Anim. Ethol.* 1:369–377.

SCHNEIRLA, T. C., J. S. ROSENBLATT, AND E. TOBACH. 1963. Maternal behavior in the cat. In *Maternal behavior in mammals*, pp. 122–168, ed. H. L. Rheingold. New York: Wiley.

SEABROOK, M. J. 1972. A study to determine the influence of the herdsman's personality on milk yield. *J. Agric. Lab. Sci.* 1:45–49.

SHILLITO, E. E. 1975. A comparison of the role of vision and hearing in lambs finding their own dams. *Appl. Anim. Ethol.* 1:369–377.

SINGH, P., AND M. HOFER. 1978. Oxytocin reinstates maternal olfactory cues for nipple orientation and attachment in rat pups. *Physiol. Behav.* 20:385–390.

SMITH, F. V., C. VAN-TOLLER, AND T. BOYES. 1966. The critical period in the attachment of lambs and ewes. *Anim. Behav.* 14:120–125.

STERN, J., AND S. LEVINE. 1972. Pituitary-adrenal activity in the post-partum rat in the absence of suckling stimulation. *Horm. Behav.* 3:237–246.

TEICHER, M., AND E. BLASS. 1976. Suckling in newborn rats: Eliminated by nipple lavage, reinstated by pup saliva. *Science* 193:422–424.

TEICHER, M., AND E. BLASS. 1977. First suckling response of the newborn albino rat: The roles of olfaction and amniotic fluid. *Science* 198:635–636.

TERKEL, J., D. A. DAMASSA, AND C. H. SAWYER. 1979. Ultrasonic cries from infant rats stimulate prolactin release in lactating mothers. *Hormone Beh.* 12:95–102.

TOMLINSON, K. A., E. O. PRICE, AND D. T. TORELL. 1982. Responses of tranquilized post-partum ewes to alien lambs. *Appl. Anim. Ethol.* 8:109–117.

TYLER, S. J. 1972. The behaviour and social organization of the New Forest ponies. *Anim. Behav. Monogr.* 5:87–196.

VOITH, V. L. 1980. Functional significance of pseudocyesis. *Mod. Vet. Pract.* 61:75–77.

WOLFF, P. 1973. Natural history of sucking patterns in infant goats: A comparative study. *J. Comp. Physiol. Psychol.* 84:252–257.

WOLSKI, T. R., K. A. HOUPT, AND R. ARONSON. 1980. The role of the senses in mare-foal recognition. *Appl. Anim. Ethol.* 6:121–138.

WOOD, P. D. P., G. F. SMITH, AND M. F. LISLE. 1967. A survey of inter-sucking in dairy herds in England and Wales. *Vet. Rec.* 81:396–398.

ZARROW, M. X., V. H. DENENBERG, AND B. W. SACHS. 1972. Hormones and maternal behavior in mammals. In *Hormones and behavior,* ed. S. Levine. New York: Academic.

Animals must eat to survive. They must take in certain quantities of energy and, depending on species-specific requirements and physiological demands, obtain vitamins, minerals, amino acids, and essential fatty acids. Our concerns with feeding behavior involve not only the species-typical behaviors that have evolved to allow animals such as prey species to obtain the required food in the face of threatened predation, but also the manner in which certain food materials are preferred or rejected. Some materials are toxic or harmful if eaten in large quantities; other foodstuffs may be deficient in some way such that if they are consumed to the exclusion of other foodstuffs, disease could result. Of special concern with regard to domestic animals is the consumption of too much or too little food, leading to obesity or anorexia.

In this chapter we shall consider the species-typical characteristics of feeding in carnivores and herbivores and take up rumination as a special topic related to feeding mechanisms in ruminants. The discussion will then focus on some of the environmental, hormonal, and physiological aspects of feeding behavior as background for understanding overeating and un-

dereating. Since eliminative behavior is related to food intake
we shall also consider this topic here.

One of the most important aspects of animal domestication
has been human control over the type and quantity of food
consumed and the conditions under which the food is eaten. In
raising and maintaining animals we tend to provide food rather
than demanding that the animals spend time hunting or grazing.
Thus it is not unusual to find behavioral problems reflecting an
increase in spare time and presumably boredom. Behaviors such
as cage pacing, harassment of other animals, and household
destructiveness (by dogs) are examples of problems that may be
caused by boredom. Economically important behavioral disturbances such as tail biting in pigs and feather pecking in poultry may be due to the fact that the animals spend such little time
each day foraging that they become bored. It has been suggested
that floor feeding of pigs may reduce tail biting by forcing the
animals to spend more time picking up the food from the floor
(Ewbank, 1969). Feeding of completely pelleted diets to livestock
and horses has become an important economic consideration
from the standpoint of ease of transportation, storage, and feeding. However, pelleted feed reduces feeding time. In horses this
reduction of feeding time leads to an increase in chewing wood
(Haenlein et al., 1966; Willard et al., 1973) (Figure 5-1).

In general, carnivores spend very little time eating, although
in the wild considerable time would be spent hunting. Their diet
is highly nutritious and concentrated. Herbivores, on the other
hand, spend approximately 30 to 40 percent of a 24-hour day
eating (Arnold and Dudzinski, 1978). This feeding pattern is related to the relatively low energy and low nutritive value of the
food that herbivores are adapted to consume. Animals classified
as omnivores, such as pigs, rats, and poultry, eat a variety of
materials. Pigs will eat roots, grasses, earthworms, slugs, snakes,
and frogs. On pasture, pigs may feed or graze for five hours or
more, but if fed concentrates, their eating time may not exceed
10 minutes (Signoret et al., 1975).

The effects of social relations among animals that live under
conditions of intensive grouping have influenced feeding behavior. Usually the most dominant animals get the best supply
of food. If the feeding facilities are too small, then subordinate
animals may get an inadequate diet. In one study of goats, it
was found that the subordinate animals were carrying a much
heavier parasite load than the more dominant individuals. In

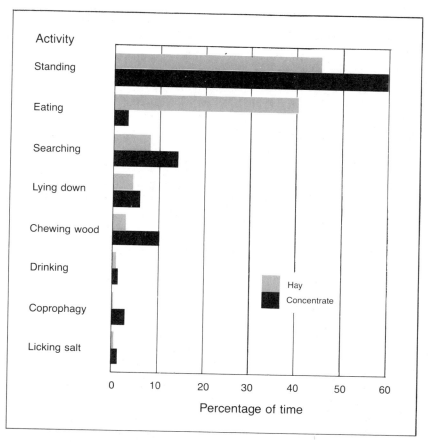

Figure 5-1 Comparison of average percent of time spent in various activities by horses fed hay or concentrate: Note the increase in chewing wood when concentrate is fed. (Data drawn from Willard et al., 1977.)

examining social relationships during feeding, it was determined that the subordinates were forced to consume food from the ground, where they undoubtedly picked up an excessive number of parasite eggs (Campbell and Fraser, 1961).

■

FEEDING BEHAVIOR OF DOGS AND CATS

Dogs, like their wild ancestors, tend to eat very rapidly, which probably reflects the fact that when prey is killed, there is competition for food, and the animals that eat the most rapidly get the most food. In Africa animals that kill prey often find themselves competing with other carnivores such as hyenas for the kill. The life of a carnivorous predator is often one of feast or famine. A wolf has a remarkable capacity to gorge itself; it has been reported that wolves can consume up to a fifth of their own body weight in meat at one sitting. Some domestic dogs can also consume enormous meals; a male Labrador was observed to consume 10 percent of its body weight in canned food at once (Mugford, 1977).

The wild ancestors of cats were solitary hunters that existed on small rodents and birds rather than on large game, as would larger cats or wolves. Thus they tended to feed frequently and on smaller meals. A mouse is about the size of a normal meal for a house cat. The difference in capacity of dogs and cats to consume large meals may be related to the fact that obesity is a greater problem in dogs than cats.

Social Influences

In many animals food consumption increases if several animals feed at one time. This social facilitation in feeding has been observed in dogs, particularly puppies, when they are fed a few times during the day. An increase of food intake of 40 to 50 percent has been reported when puppies were fed in groups compared with intake when fed alone (Ross and Ross, 1949a, 1949b). Such social facilitation has not been observed in cats.

The effects of social facilitation are temporary, however, and experimental observations show that over the long run group-fed puppies probably do not eat more than those fed alone. Social facilitation is less apparent when food is readily available all day. In experiments in which adult miniature poodles and adult cats had continuous access to complete semimoist dog food, there was little evidence of social facilitation in either the dogs or the cats (Mugford, 1977). Both species tended to take a large number of small meals. The poodles tended to eat mostly at night, and the cats ate throughout the 24 hours.

When feeding is restricted, however, the food dish is one area where dogs, especially puppies, tend to define their dominance relationships. The dominant animals may get such a large proportion of the food that the more subordinate ones are undernourished. This, of course, can be prevented by feeding puppies with several pans of food.

Types of Food

Although it is customary to think that a natural diet for carnivores consists of meat, to believe that these animals are perfectly capable of living on meat alone is misleading. This misconception overlooks the fact that the diet of feral, or wild, canids and felids consists of the entire carcass of small birds, rodents, and larger mammals as well as occasional vegetative matter. Thus feathers, skin, bones, and the contents of the intestinal tract of herbivorous birds and mammals constitute a good part of the diet.

Dogs are the only domestic species exhibiting a behavioral pattern that could be called hoarding or food storage, a trait that is typical of wild canids. A common example is the tendency of some dogs to bury bones. The behavior is sometimes so strong that even when dogs are in plastic or metal cages they will sometimes attempt to cover a bone using papers on the cage floor.

Burying small prey has been observed in wild canids, including wolves and the red fox (Henry, 1977; Jeselnik and Brisbin, 1980). These animals have been observed to go back to the source of the cache during lean months of the year. The famous ethologist Tinbergen (1972) observed interactions between gulls and the European red fox on a sandy beach, and found that a fox would carry gull eggs and place them in a small hole it had dug with its forepaws. An egg was covered by the fox using its nose to push sand over the hole. Several eggs were buried in different scattered locations on the same day. The eggs were dug up and eaten later in the year when the gulls had gone and food was scarce. The eggs were located by the fox first going to the general area where they had been buried and then finding the particular spot by smelling the eggs.

It is common for dogs to bury bones, but not to dig them up later. Some dog owners claim to have seen their animals bury small animals such as woodchucks and later uncover them.

Both dogs and cats have been observed to eat grass, especially long blades. This phenomenon, which has been observed in wild carnivores, has not been studied experimentally, and so nothing can be said conclusively about the reasons for eating grass or the physiological effects. Cat and dog owners sometimes claim that animals tend to eat grass when they have a gastroenteric disorder. Often they vomit after eating the grass, suggesting that irritation of the gastric mucosa by the grass induces the regurgitation. Other pet owners also report that eating grass is followed by a loose bowel movement, suggesting that grass has a laxative effect. One might argue that the feather, fur, bones, and so forth consumed in a diet of small game has a constipating effect that is counteracted by eating grass. Another hypothesis that might be explored is that long blades of grass have the mechanical effect of removing parasites from the intestines.

Pica: Ingestion of Unnatural Objects

Carnivores are occasionally observed to chew and ingest unnatural objects such as gravel or pieces of wood. There is no explanation for this behavior other than possibly the rewarding attention such animals may receive from their owners. Eating feces is referred to as coprophagy. In jackals coprophagy has been observed, and it is common in mothers who consume the waste products of their infants. Aside from this, the behavior is apparently abnormal in carnivores. Coprophagy does occur in dogs, possibly as a behavior related to social boredom and sometimes as an attention-getting act.

An interesting aberration of ingestive behavior is wool chewing in cats. This is a behavioral trait that seems to be restricted to the Siamese breed. At about the time of puberty cats may suck, chew, and ingest chunks of wool from stockings, sweaters, or caps. Wool is usually sought first, but many cats have generalized to other types of cloth and even synthetic fibers. Most cats seem to pass through their wool-chewing phase in a year or two. Others threaten to ruin an entire wardrobe and cannot be kept as indoor pets. As for other forms of pica, there is no widely accepted explanation for this behavior (Hart, 1976).

■

FEEDING BEHAVIOR OF HERBIVORES

The intestinal tract of the herbivore is highly adapted to utilizing plant material high in cellulose and low in nutritive value. A great deal of this low-energy food must be consumed to satisfy the physiological demands of herbivores. The intestinal tract is correspondingly greatly enlarged over that of carnivores. Ruminants have a large complex stomach, of which the rumen is the largest part and in which bacterial action breaks down plant cellulose and other plant materials into short-chain fatty acids and other products that are abosrbed through the intestinal tract. Horses have a simple stomach, but the colon and cecum are very large and serve as cavities for bacterial action similar to that occurring in the rumen of ruminants.

Grazing and Foraging

Grazing in ruminants in temperate climates on good pasture usually occurs in four to five periods per day, with the total cumulative amount of time grazing being from four to nine hours per day. Horses may spend about twice as long grazing as ruminants (Boy and Duncan, 1979). The total daily intake of forage is about 10 percent of body weight (Hafez and Bouissou, 1975; Tribe, 1955). In cattle the major periods of grazing are at sunrise and sunset, with some grazing also occurring at mid-morning and early afternoon. Grazing activity is alternated with exploring, resting, and ruminating. In a herd of animals one usually sees a good deal of social facilitation in that all of the animals tend to engage in the same activities at the same time. Thus, cows tend to all graze at one time or all ruminate or rest at one time.

The type or quality of feedstuff can dramatically affect feeding behavior. Feeding concentrates reduces feeding time. In poorer pastures cattle graze for much longer periods and travel greater distances. In some instances rumination may not occur at all during daytime grazing (Figure 5-2) (Lott and Hart, 1979).

Free-ranging sheep characteristically start to graze just before sunrise and stop sometime after midmorning. They rest and then take up grazing in midafternoon and continue until sometime after sundown. The long period between the morning and afternoon grazing is devoted to idling and rumination (Squires, 1971). Whether a day is overcast or rainy does not seem to have much effect on a sheep's grazing (Bueno and Ruckebusch, 1979).

Temperature and humidity can affect the grazing pattern. Sheep do not like to graze in the middle of a hot day. Some investigators have mentioned that sheep can sense whether the day is going to be hot before the sun comes up, and they start grazing sooner in the morning and stop earlier on those days (Dudzinski and Arnold, 1979).

Figure 5-2 On poor pasture, such as this in the northern Nigeria savanna, cattle often spend the entire day grazing and ruminate only at night.

Of all livestock species, sheep tend to graze in the most clustered formation, feeding in groups. This behavior is attributed to the fact that this species historically has been under heavy predation pressure from wolves, jackals, hyenas, and dingoes, and the best survival strategy for individuals is to graze as part of a group (Hamilton, 1971). However the avoidance of predators by grazing in a group works best when the pasture is lush. As a pasture is eaten down and forage conditions decline, the flocking pattern in sheep grazing can break down as the sheep disperse, grazing in smaller flocks (Figure 5-3). The relation of dispersal pattern to forage conditions was so accurate in one study in Australia that forage conditions could be "calibrated" by aerial photography of flock size (Dudzinski et al., 1978).

The impressive adaptability of herbivores in utilizing grasses and plants does not mean that these animals indiscriminately consume whatever material is available. If this were the case, one would predict that grazing behavior would reflect the most efficient harvesting mechanism. However, if one just looks at grazing as a mechanical harvesting procedure, predicted movements do not correspond to the actual observed grazing behavior. The conclusion is that herbivores are not acting like harvesters, but foraging quite selectively and utilizing sensory

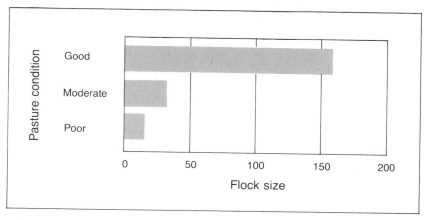

Figure 5-3 Sheep tend to graze in large flocks when forage is plentiful. When forage availability becomes poor, sheep graze in much smaller flocks. This, of course, allows them better access to forage, but it does increase their exposure to predators. (Data from Dudzinski et al., 1978).

input to guide their behavior (Pyke, 1978). Cattle and horses will avoid brush and most plant leaves in favor of grass. This selection is related to the bitter taste of shrubs and brush foliage. Most toxic plants are avoided by herbivores, and work with the guinea pig, which is a herbivore, indicates that this avoidance may be due in part to an aversion acquired as a result of once having suffered a gastrointestinal illness from toxic plants (Jacobs and Labows, 1979).

Goats can tolerate substances with a much more bitter taste than cattle and horses and often browse leaves of shrubs, small trees, and brush (Bell, 1959). In a pasture that is overstocked, goats may do a good deal of permanent damage because the few plants left by other herbivores are destroyed. In some of the sub-Saharan regions of Africa overgrazing by cattle has eliminated pasture suitable for cattle grazing, so that only goats can survive. However, goats can completely eliminate all vegetation, and this is believed to be one of the main factors that led to the formation of the desert thousands of years ago and that is currently responsible for the advancement of the desert several kilometers each year.

Even on grassy pasture not all areas will be grazed evenly by herbivores. Usually the younger and most succulent plants are eaten first. Some species of grasses or small plants are preferred to others (Fontenot and Blaser, 1965). Thus a pasture is selectively grazed by a type of creaming of the most favored plants and progressive defoliation of the most palatable parts of the plants. By selective grazing, animals consume forage higher in protein and lower in crude fiber than could be obtained by indiscriminate harvesting by a farmer (Wier and Torell, 1959).

Of course, as the pasture is eaten down the animals become less selective, and the quality of the forage consumed is reduced. When the forage is plentiful, most herbivores will avoid grazing areas that have been recently contaminated by feces or urine. The aversion appears to be due to the odor of feces rather than to changes in the palatability of the vegetation growing in soiled areas (Ödberg, 1975). Due to avoidance of contaminated forage, it is not unusual to see small stands of lush growth in a pasture surrounded by closely cropped grass.

This avoidance behavior obviously plays an important role in reducing the ingestion of eggs of internal parasites that are passed in feces. The tendency to avoid feces is strong enough that nomadic herdsmen have taken advantage of this behavior in calves to wean them from their mother by applying fresh feces to a cow's udder (Lott and Hart, 1979). When pasture is overgrazed, herbivores become less selective regarding contaminated forage and are thus more likely to ingest parasite eggs as well as less nutritious food.

Some of the preference of herbivores for certain plants seems to reflect a familiarity with certain plants and establishment of habit in foraging. Cattle raised in certain areas of South Africa become browsers, feeding on leaves of shrubs and brush, because of the lack of grass. When transferred to grassland areas, they may take months to adapt not only in terms of forage eaten but in changing from browsers to grazers (Ewer, 1968).

Rumination

An important aspect of ingestive activity in cattle, sheep, and goats is rumination, which consists of regurgitating, remasticating, and reswallowing slightly chewed vegetative matter that has been previously taken into the rumen. Wild ruminants make maximal use of the storage capacity of their complex stomach by taking in unchewed grass relatively quickly in an open field where they can watch for predators. The plant material is then usually chewed more completely during rumination in quieter and safer surroundings.

Rumination periods ranging from two minutes to an hour or longer occur many times throughout the day and are usually interspersed with grazing, feeding, resting, and exploration. Estimates of rumination periods range from about eight to 20 per day. Most rumination occurs at night. The total amount of time spent ruminating varies depending on the species and the type of forage eaten. Cattle spend about as much time ruminating as grazing, whereas sheep about half as much time ruminating as grazing in the same pasture (Lofgren, 1957). Apparently, since sheep select less fibrous herbage, take smaller bites, and swallow finer material, the regurgitated food requires less chewing (Arnold and Dudzinski, 1978). Generally, with high-quality pasture the proportion of time spent grazing is longer, and ruminating is shorter (Tribe, 1955). It is possible for animals to ruminate while standing, but they usually lie down, with the chest against the ground. During illness or estrus, rumination time is reduced.

Ruminants spend much less time sleeping than people, dogs, and horses. Whereas most nonruminants sleep on their sides, it is necessary for the ruminant to maintain an upright posture so that the esophageal opening to the rumen does not drop below the fluid level, which would prevent belching of rumen gases and cause bloat. Therefore, ruminants only sleep for a few minutes at a time.

Since adult ruminants do not sleep for a prolonged period of time the way animals with simple stomachs do, it has been proposed that rumination in some way provides the physiological rest and rejuvenation of deep sleep (Bell, 1960; Bell and Lawn, 1957). In fact, there is a striking resemblance between rumination and several characteristics of sleep. Rumination occurs in quiet surroundings, mostly at night. The animal goes into rumination behavior relatively slowly, which is characteristic of an animal falling asleep, and environmental stimuli rapidly arouse the animal from rumination, much as intense noises arouse a sleeping nonruminant.

Some of the earlier work involving electroencephalographic recording from ruminants during rumination suggested that the brain-wave pattern characteristic of deep sleep in nonruminants occurred in ruminants during rumination. Now, however, it is evident that there is no specific type of brain wave during rumination (Bell and Itabisashi, 1973).

■

FOOD PREFERENCES AND AVERSIONS

Although limitations on food intake and certain preferences or aversions in animals are genetically determined, different individual animals show various food preferences depending on their early experience with food, nutritional state, and possible adverse experiences with food.

One of the strongest determinants of food preference is the degree of hunger. As food is ingested and an individual becomes somewhat satiated, material that might previously have been eaten may no longer be acceptable and only the most palatable or favored foods may be selected. The starved or very hungry animal may readily eat foods that are usually marginal in palatability.

Effects of Early Experience with Certain Foods

In most mammalian species, young are assisted in learning to obtain appropriate food, usually by direct or indirect tutelage from the mother, and sometimes the father (Geyer and Kare, 1981). However, studies of the relationship between early food experience and preferences in adulthood are scanty.

Many dog and even cat owners are aware of the fact that their pets will eat vegetables and other "unnatural" foods as adults if they are introduced to these foods as young animals. One of the more extreme claims of the development of food preferences as a function of early experience was made by Kuo (1967). He reports an experiment in which he obtained chow pups just after birth and hand-fed them different diets first in the form of a liquid food, then semisolid food, and later solid food. To one group he fed only soybeans, to another group only fruits and vegetables, and to a third group a large assortment of food including milk, butter, cheese, different types of meat, fish, fruits, and vegetables. All the diets were enriched with vitamins, minerals, and salt. At six months of age the dogs were tested for their response to different foods following 24 hours of food deprivation. Almost all of the dogs in the group initially fed soybeans refused to touch any new food and none of the animals in this group ate any meat, fish, eggs, or milk. Animals in the group that was fed fruits and vegetables would not touch any animal protein such as meat, fish, or eggs, and only a few animals ate any new food at all. Dogs in the group that was fed a large variety of foods ate virtually any type of new food, except that which was bitter, sour, or stale. A similar experiment conducted with kittens resulted in the same pattern of food acceptance.

These results were challenged by some recent work in which puppies and kittens were fed either of two commercial diets from the time of weaning for the next 16 weeks (Mugford, 1977). When later allowed a choice between the two diets, both dogs and cats went for the novel food and not the one they had been fed.

The experiments by Mugford did not include feeding a diet from the day of birth, and the diets were not as pure in terms of being strictly animal or plant ingredients, as in the experiments by Kuo. Nonetheless, they are more along the lines of practical application. Mugford stresses that feeding a single food mixture, even of a balanced diet, induces a transient depression of its relative palatability. In fact, Mugford presents data to show that a cat's caloric intake can be markedly increased by feeding a succession of different diets and allowing cats to feed three times a day.

There has been limited work in examining the effects of early experience with a food on later preferences in herbivores. This is an area that might receive more attention as we attempt to find inexpensive foodstuffs for livestock, some of which may be quite nutritious but would not be palatable to the animals. Authorities have noted that since ruminants (or rather their bacteria) can digest cellulose, we could convert old newspapers into milk or beefsteaks. Such use of paper or crop residues requires a supplementation of nitrogen, usually in the form of urea. Farmers have tried to induce their animals to lick urea blocks by mixing the urea with molasses. The results are variable. In sheep, for example, almost all the members of one flock may lick the block whereas no members of another flock touch it. However, offering molasses-urea blocks to lambs before weaning establishes an acceptance or preference for the material, and nearly all such lambs will consume some of the blocks later. Without the preweaning exposure older lambs are not likely to touch the blocks (Lobato et al., 1980). The same effect has been found for acceptance of other foodstuffs. For example, adult sheep accept grain more readily when as young they had been exposed to grain and had observed adults consuming it. The acquisition of a taste for supplements, such as molasses-urea blocks, may be a result of lambs imitating their mother's feeding behavior or from exposure to the taste through the mother's milk.

Another example of the establishment of taste preferences early in life is the observation that when sheep consume a wide range of plant species early in life, they succeed at foraging better on new and unfamiliar pastures later (Arnold and Maller, 1977).

Specific Hungers

Specific hungers for several minerals, vitamins, and amino acids have been claimed for different animals. The best-known example is the specific hunger for sodium, which can be induced in rats or other animals by removing the adrenal glands. Adrenalectomy results in rapid elimination of the sodium ion, and animals so treated will show a strong preference for substances containing sodium immediately on tasting it. Such animals will also have a strong preference for similar-tasting salts that do not actually alleviate their deficiency (Denton, 1967; Nachman, 1962). Sodium-deficient sheep, goats, cattle, and other animals can distinguish sodium in foods and can even adjust their intake of the ion by consuming different amounts of varying concentrations (Blair-West et al., 1968; Denton, 1967).

Rats that have been made deficient in calcium, sodium, and magnesium, the vitamins thiamine, pyridoxine, and riboflavin, or the amino acid histidine have preferences for foods containing these substances (Rozin and Rodgers, 1967; Rogers and Harper, 1970).

Dietary Self-Selection   Much of the initial work and theoretical thinking on specific hungers must be attributed to Richter and his concept of the importance of specific hungers in maintaining a nutritional homeostatic condition. Probably the best-known demonstration of individual vitamin and mineral specific hungers is his cafeteria self-selection experiment in which rats apparently selected an extremely well-balanced diet from a large number of food trays (Richter, 1943). Since the earlier work of Richter, numerous investigations have been designed to examine the ability of animals to select a diet that provides for adequate or optimal health. Studies of this "nutritional wisdom" concept in rats have shown that the animals have a surprising ability to consume a balanced diet (Overmann, 1976).

Animals are also able to adjust their diets according to changing needs. Rats housed in the cold increase their total caloric intake relative to controls by choosing more carbohydrate (Leshner et al., 1971). Rats that are made underweight show a more efficient food utilization than do control animals (Boyle et al., 1978).

Of the domestic animals, sheep, swine, and cattle have also been reported to make selective dietary choices. For example, cows allowed to graze from four haystacks were observed to feed exclusively from the single stack made up of hay grown on fertilized soil (Albright, 1945). In a more recent study, dairy cows provided with free-choice minerals and palatable hay selected a diet that came close to dietary needs (Muller et al., 1977).

An almost equal number of studies illustrate that the capabilities of animals to select an adequate diet are limited. In one experiment pregnant ewes did not select a diet sufficiently high in protein to maintain normal pregnancy (Gordon and Tribe, 1951). Animals may not always wisely choose the foodstuffs they need partially because of a relaxation in natural selection pressures: in many instances, we have taken food selection almost completely out of their control. Self-selection of diet no longer pays off. Our knowledge of nutrition allows us to supplement diets with the necessary vitamins, minerals, and amino acids. If an animal wanted to display its own innate wisdom about nutrient selection, it would have little opportunity.

Specific Hungers versus Postingestional Effects   There are two ways of looking at the way animals are able to select diets according to their nutritional status. One is by assuming genetic programming for specific taste preferences or for a range of taste preferences for nutritious foods. A mechanism that pinpointed only specific food tastes would be somewhat biologically uneconomical, because most of the specific hungers would not be utilized in an animal's lifetime, but a range of inherited taste preferences might be useful. For example, the widespread preference for sweet taste would facilitate consumption of energy-rich carbohydrates, whereas the common aversion to bitter taste would often protect the animal from toxic substances.

Another possible mechanism suggested by investigators working in the area is that animals learn, through postingestional effects, what makes them feel better when they are deficient in something. An animal might, in its lifetime, come to learn preferences for the tastes of foods associated with recovery from deficiencies it experienced. Most preferences that animals show for a food that contains a substance they are deficient in are acquired over several exposures to the food (Rozin and Kalat, 1971).

Most specific-hunger experiments have pitted the old food against a new or novel one in a two-choice preference situation. When a deficiency has been created by a food, it is argued that the choice of the new, complete food represents an aversion to the old, deficient food. In other words, it is not that the animal "loves the new stuff" but rather "hates the old stuff." It has been observed, for example, that rats fed a diet deficient in thiamine exhibited a number of behavioral patterns revealing aversion to the diet. As the deficiency developed, less time was spent eating, and the animals began spilling their food, after first sniffing it, by scooping the food out of the tray so that it fell through the wire mesh floor. Sometimes redirected feeding was observed in which the rats would first investigate the food cup and then move over to the wooden barrier separating the nest area from the rest of the cage and begin to chew it vigorously.

Even if the rats were given a vitamin-enriched diet and allowed to recover from their deficiency, they still responded to the old deficient diet as though it were aversive. This indicates that the nausea and other ill effects produced by ingestion of the deficient diet lead to a learned aversion to the familiar deficient food. A new food that does not produce nausea is therefore preferred.

Amino acid imbalance and deficiency also causes depression in food intake (Rogers and Leung, 1977). Feeding corn meal and bone meal to pigs results in a depression of food intake presumably due to a deficiency of the amino acid tryptophan. This is reversed by supplementing the diet with tryptophan (Montgomery et al., 1978).

The model of learned aversions as a basis of apparent specific hungers is an appealing alternative to the theory that animals have a multitude of genetically determined specific hungers. With the learning approach an organism can deal with a variable and changing environment where programmed and specific modes of response may be maladaptive.

Specific Hunger for Sodium   Of all the nutrients in the diet, sodium appears to be the most critical because of its importance in fluid balance, the relatively large amount needed, and its common scarcity in the environment. To meet these particular physiological needs, Rozin and Kalat suggest that this is the one nutrient for which there has been a selection pressure for a genetically programmed specific hunger. Animals deficient in sodium show a strong preference for sodium immediately on tasting it even when given a wide choice of new diets. As I have pointed out, this is not really the case with regard to other deficiencies.

Coprophagy   Animals of several species, especially the young, engage in occasional coprophagy in which they consume feces of their own or other species. Among domestic animals, this has been noted in foals. Feces of adult animals contain vitamins, minerals, and beneficial bacteria that may help establish an appropriate intestinal flora in young animals. Coprophagy that leads to the establishment of normal intestinal flora thus appears to be due to inherited behavioral predispositions.

Taste Aversions

It should come as no surprise to anyone who has unintentionally eaten spoiled food that animals may develop an aversion to a food that has made them sick or nauseous. In the laboratory there are examples of acquired aversions to foods that have been paired with sublethal doses of x-rays, lithium, apomorphine, and some psychoactive drugs including methylphenidate (Ritalin), which is used to treat hyperactivity in children and occasionally in dogs (Riley and Zellner, 1978). If these various treatments cause postingestional nausea or sickness, then the taste or smell of the food associated with the treatments takes on aversive properties after one or more pairings with the illness-producing treatment (Garcia et al., 1974). An example of an aversion to meat in cats produced by poisoning the meat with lithium chloride is shown in Figure 5-4.

Conditioning of aversions is commonly experienced among people. Almost everyone can remember a food that they pre-

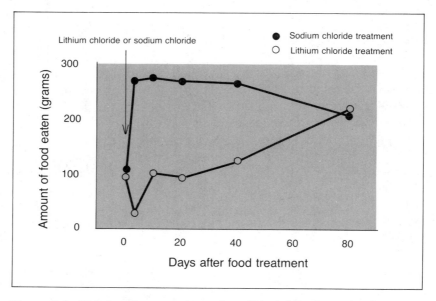

Figure 5-4   This is a demonstration of conditioned food aversion in cats. Animals were maintained on dry food until day 0. At that time a very palatable meat diet was offered, while some cats were treated with lithium chloride and the others with sodium chloride. At 3, 10, 20, 40, and 80 days after treatment the cats were offered the meat diet without treatment. The aversion in cats given just one exposure to lithium chloride was maintained for 40 days. (From Mugford, 1977.)

viously favored but that later became aversive after the food was associated with a gastrointestinal illness and nausea. Human cancer patients suffering from unpleasant side effects of chemotherapy or radiotherapy sometimes acquire aversions to the food they have eaten just before the treatment (Bernstein, 1978). Even the aversive physiological effects of a tumor itself may induce an acquired aversion to the food consumed during the growth of the tumor. The latter effect has been documented in rats (Bernstein and Sigmundi, 1980). Conditioning of aversions is also involved in the phenomenon in which various populations of rats have developed a bait shyness to poison bait that previously was effective in reducing the rat population.

Coyote Predation Control   One of the more enterprising applications of conditioning aversions is that involving an attempt to produce an aversion in coyotes to sheep as a possible approach to the control of sheep predation (Gustavson et al., 1974). The coyotes were known to attack sheep even in captivity. The investigators wrapped lamb meat laced with lithium chloride in a lamb wool covering and allowed the coyotes to eat it. After a couple of experiences with the poisoned lamb meat the coyotes no longer chased sheep in the experimental enclosure and, in fact, the investigators related seeing one of their coyotes being chased around the pen by a lamb. The specificity of the conditioning is illustrated by the fact that the coyotes would readily go after rabbits even though they reportedly had an aversion to sheep.

The use of lithium-laced poison bait to reduce coyote and wolf attacks on sheep seems to have had limited success in field trials (Gustavson et al., 1976; Ellins et al., 1977). After the coyote or wolf feeds on the bait it becomes ill. Later, when it has access to live prey, the animal may attack and kill the prey but does not eat it. At this time an association appears to be made between the smell and the taste of the prey, and the predator subsequently acts as though the smell were also aversive (Gustavson et al., 1976).

Conditioning of Food Aversions   How can a learned food aversion be established when the interval between eating the food and the appearance of postingestional illness may range from up to 15 to 60 minutes? Those who have advanced the concept of conditioned food aversions have proposed that there is a specific, long-delay learning system that operates to meet the peculiar demands of feeding behavior that is not found in other types of learning. It is assumed that there is a natural predisposition for mammals to associate food tastes, rather than visual cues, with postingestional stimuli. Conditioned aversions are most easily produced to foods that are somewhat novel; familiar tastes seem to be quite resistant to association with illness. Thus, aversions to poisonous compounds used in pest control may be formed when animals ingest nonlethal doses of novel-tasting baits. The ability to associate a food with gastrointestinal sickness is diminished once the animal makes the association that the flavor is safe.

The function of conditioned food aversions in nature is to protect animals from repeated ingestion of food that produces gastrointestinal illness. Foods that are likely to contain an endotoxin will thus be avoided. In situations where animals develop food allergies, a conditioned aversion would lead to avoidance of the allergy-producing food. Conditioned food aversions also play an important role in pest control, since animals that ingest sub-lethal doses of a toxic bait will subsequently avoid it. These food aversions can be socially transmitted to the young (Galef and Clark, 1971).

Aversion to Cannibalism   An interesting food aversion found in carnivores is a reluctance to feed on dead carcasses of mem-bers of their own species, even if very hungry. From one stand-point conspecific flesh should make an ideal food because it contains the same materials as the feeder. The aversion to can-nibalism is, however, widespread among carnivores (Fox, 1975). A biological explanation of this behavior is that it may be a mech-anism to preclude the intake of disease organisms that killed the conspecific animal in the first place. In rats this aversion to can-nibalism is tied to the smell and taste of the skin and fur. Skinned carcasses of conspecifics are readily eaten (Carr et al., 1979).

■

CONTROL OF FOOD INTAKE

Most animals have a remarkable ability to adjust their con-sumption of calories in terms of carbohydrates, fats, and proteins to meet their energy requirements and maintain a normal body weight. Furthermore, animals that are force-fed a partial diet or given calorically diluted food adjust their food consumption ac-cordingly. Rats maintained in an environment of low temper-ature respond by voluntarily increasing energy intake, which tends to compensate for the energy expenditure necessary to maintain body temperature (Hamilton, 1967). Livestock species living outdoors in the winter probably do the same.

An interesting illustration of normal weight control is the observation that for people an average daily error in consump-tion of as much as a handful of peanuts is sufficient to add a few pounds of fat tissue per year. Most people, of course, reg-ulate their weight within a much finer degree than this.

The mechanisms by which animals regulate the amount of food they consume during each meal, or on a long-term basis, has been the object of much research. Among the factors that have been considered are stimuli concerned with the passage of food through the mouth and other parts of the digestive tract, distention of the stomach, osmotic changes due to body tissue water being sequestered by the gut, the release of substances such as glucose, amino acids, or fatty acids into the blood where they are detected by brain chemoreceptors, and the release of humoral factors indicating satiety. None of these factors acting alone seems to be responsible for the control of satiety or appetite, but all of them probably contribute to the control (Panskepp, 1971).

In particular, the experiments pointing to humoral factors may prove quite important in understanding the control of feeding behavior. An illustration of this principle is an experiment in sheep showing that when blood from a hungry and a satiated sheep was cross-circulated, food consumption in the hungry sheep was depressed and food consumption in the satiated sheep was increased (Seoane et al., 1972).

The release of the gut hormone cholecystokinin from the duodenum and the action of this hormone on the brain is a postulated mechanism of satiety that is gaining favor (Della-Fera and Baile, 1979).

Interestingly, most mechanisms proposed for food intake control acknowledge that animals feel satiated and stop eating before a significant amount of nutrient absorption has occurred. One exception to this view is the glucostatic theory. This theory proposes that as blood glucose levels fall, an animal becomes hungry. After feeding, blood glucose increases, and glucose stimulates areas of the brain, especially the ventromedial hypothalamus, that inhibit eating. The ventromedial hypothalamus is often referred to as the satiety center. The lateral hypothalamus is considered a counterpart to the ventromedial hypothalamus and is classically referred to as the eating center. However, several objections to the classical view of the role of the hypothalamus in behavior have been raised (Friedman and Strick, 1976; Grossman, 1975). There are difficulties with the glucostatic and with other nutrient-specific theories as well. The future will probably see an array of new concepts and theories related to the physiological control of food intake.

Physiological Control of Fat Deposition: Obesity

Given the variety of short-term and long-term controls over feeding behavior that exists, it is not surprising that we rarely see permanently obese animals in nature. Of course, hibernating animals require fat stores for the long winter and marine mammals such as elephant seals and sea lions need blubber for insulation or fat stores for a long period of territorial guarding on land. Ungulates in temperate zones build up fat stores for the rut and winter. Aside from these special circumstances, excessive fat is disadvantageous for the herbivores, which have to escape from predators, and the carnivores, which must catch the fast-moving herbivores. On the domestic scene we have allowed pet dogs and cats that are too fat to survive in the wild to thrive, and we have eliminated the selection pressures maintaining close controls on food intake. With livestock animals we have been more intentional in our manipulation of food intake behavior. Dairy cattle are skinny because we breed them to put all their food energy into milk. Beef cattle are fat because we want the calories to go into marbled fat and not milk. The same goes for sheep and pigs.

To study the genetic aspects of obesity scientists have come up with the genetically obese "Zucker" rat and compared the feeding behavior of these obese rats with normal lean animals. Obese Zucker rats are the same as pigs in that they are polite enough in their eating habits, and in fact eat more slowly than their lean counterparts, but they tend to eat all day long (Wangsness et al., 1980). The obesity in genetically obese pigs is unquestionably unhealthy; these animals have excessive fat deposition, inferior muscle development, and a reduced growth rate (Cote and Wangsness, 1978).

Some recent work on obese rats points to a possible biochemical mechanism by which the inherited tendency to be obese occurs. The animals possess an excess of the naturally occurring brain opiate β endorphin in the pituitary gland (Margules et al., 1978). Interestingly, these opiates may be increased in the brain by behaviorally stressful situations, suggesting that this mechanism may be involved in stress-induced obesity in humans (Morley and Levine, 1980).

One of the controversial areas of obesity research has to do with the relationship of obesity to an individual's number of fat cells. Some evidence has indicated that the number of fat cells possessed by an individual is relatively fixed soon after birth. If an animal has an excessive number of fat cells at birth, it has a predisposition to become fat later in life. The number of fat cells is regulated by genetic influences. Also, overfeeding in infancy may increase the number of cells. There is some tendency for the brain to "defend" the fat cells: an animal that is obese, but forced to reduce to the weight of a normal animal, may behave in the same way as a normal animal that is food deprived. It is more irritable, less active, and shows reduced sexual activity (Hirsch and Han, 1969; Hirsch and Knittle, 1970; Nisbett, 1972).

Obesity in Domestic Animals   The degree of obesity in animals, and people too for that matter, is obviously related to causal factors besides a genetic predisposition. One of these is dietary— too much of a good thing. Human ethologists make a great deal out of the point that we are programmed to like the taste of sugar. In the days of the subsistence hunting and gathering life style we avidly ate up all the sugary fruits we found and took in the needed calories. Sugary fruits were few and far between, so that early humans never got fat. This innate predilection for sugary fruits gets us into trouble when there is an endless supply in candy vending machines.

A parallel situation occurs in dogs given highly palatable meat by-products. Their ancestors were capable of consuming a three-day supply at one feeding. This is a reflection of the feast-or-famine existence of wolves; prey are not continuously available. Dogs can eat all they need for one day and still eagerly attack the next bowl of food, and we think since they are hungry they have not had enough. This, coupled with a reduced level of exercising, is probably the main cause of obesity in dogs.

Anorexias

What makes an animal stop eating or eat very little and lose weight even in the presence of perfectly good food? We can look to nature to find adaptive reasons for such behavior as we can for almost all departures from the general pattern of nutritional homeostasis. Some appropriate examples of animal anorexias are associated with going into hibernation, migrating, or incubating eggs. The broadly stated function of these forms of anorexia is that foraging for food or eating would compete with more important survival or reproductive activities (Mrosovsky and Sherry, 1980). For a bird to leave a nest and feed would endanger the viability of its eggs. For a ground squirrel to eat during one of its occasional winter excursions to the outside would be counterproductive because its kidneys and digestive system (which were also under hibernation) would have to be geared up, and it would soon have to leave the den again for a cold trip to the "bathroom".

Normally, an animal's body weight has a certain set point, and when its weight falls below that level, the animal's appetite is stimulated. Does the hen sitting on a nest and not eating or a hibernating animal get hungrier and hungrier? Probably not, at least according to Mrosovsky and Sherry, who feel that a more reasonable explanation is that their bodies lower the set point for body fat. Thus, they do not feel hungry in these special circumstances even though they might have lost considerable weight. One factor that helps is that they tend to store fat before going into the fasting state, and most of the body weight lost is fat.

This explanation of a naturally programmed anorexia is the converse of the set-point theory of obesity outlined by Nisbett (1972) and discussed above. In obesity the set point is high, and the animal's body tends to defend a weight above normal. In anorexia the set point is low, and the body tends to defend a lower weight. Presumably the genetically obese animal is feeling constantly hungry when it is forced to reduce toward normal weight, and the hibernating animal is feeling satiated when its weight is below normal.

A form of anorexia we may see in domestic animals is that associated with fever or illness. Before explaining this some background information is needed. Animal bodies (including ours) can mobilize a variety of marvelous inflammatory and immunological processes that help fight bacterial or viral infections. These processes require energy, elevated body temperature, and inactivity so that the blood vascular and lymphatic systems can be used to greatest advantage. A long trek through the woods looking for and capturing prey or an entire day spent foraging would increase heat loss and divert blood from the site of infection (where it is needed) to the large leg muscles. Hence, a programmed loss of appetite, with a lower set point for body weight, makes it easier for the animal to lie still. As soon as the body is on the mend, and body resources can be spared for muscular activity, the appetite is back.

Just as there are several factors involved in obesity, so are there several other causes of anorexia. A specific anorexia to a particular food or class of foods, stemming from a conditioned taste aversion caused by poisoned food, was discussed above. Animals can be allergic to certain foods and will vomit, develop diarrhea, and feel very distressed after consuming the food. A common allergy in dogs is to the gluten proteins found in wheat. When repeatedly exposed to a variety of dog foods, all containing the allergen, the dog can acquire a generalized conditioned aversion to just about all commercial foods. Thus, without knowing about the allergy, we feel the dog simply has no appetite.

Psychologically related anorexias are epitomized in humans by anorexia nervosa. The only behaviorally induced anorexia in animals that might have a human counterpart is attention-getting anorexia. If a dog is off its feed we may attempt to hand-feed it, coax it along, or pamper it with special treats, thus rewarding the anorexia. The cure for animals that have learned this type of anorexia is to turn the tables on them and make eating pay off by ignoring the anorexialike behavior.

Influence on Gonadal Hormones

Although their effects are overplayed in both the human and the animal realm, hormones can influence feeding behavior and either stimulate or depress appetite. The most relevant hormones for domestic animal feeding behavior are those secreted by the gonads. Differences in gonadal hormone secretion between males and females make for sexually dimorphic differences in feeding behavior. Males eat more and weigh more than females and generally show only minor, irregular day-to-day fluctuations in eating and general activity. For example, male rats generally consume a greater portion of their diets as protein and have a greater proportion of body weight as fat (Leshner and Collier, 1973). Females undergo pronounced changes in body weight, eating behavior, and general activity in relation to the estrous cycle. One of the more interesting questions regarding hormonal control of feeding behavior and body weight is the effects of gonadectomy. It is generally felt that dogs and cats that are spayed or castrated have a higher tendency to become fat and lazy than animals with intact gonads.

Experimental work on the laboratory rat reveals that ovariectomy leads to an increase in body weight by as much as 25 percent along with an increase in food intake (Figure 5-5). The increase in weight gain is due to an increased food intake, but changes in energy balance also play a role in the weight gain, and much of the increase in weight is due to fat (Landau and Zucker, 1976; Leshner and Collier, 1973). The administration

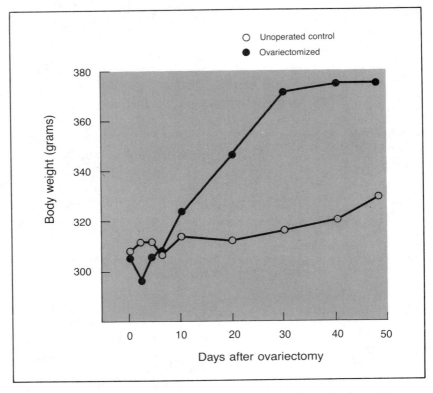

Figure 5-5 The effect of estrogen in maintaining a lower body weight is evident in this graph from a study on female rats. Body weight was recorded for 48 days after removal of the ovaries (ovariectomy), which are the main source of estrogen secretion. Both groups of animals had unlimited access to food. (From Landau and Zucker, 1976.)

of an estrogen reverses this effect and results in a reduction in food intake and body weight.

The effects of estrogen on food intake and body weight seem quite universal among animal species (Wade, 1976). For example, in ruminants during estrus and late pregnancy, when there is an increase in estrogen secretion, food intake decreases (Forbes, 1971; Tarttelin, 1968).

To what degree does experimental work, especially on rats, support the notion that spaying pet animals makes them fat? One problem with this extrapolation is that the female rat continuously cycles throughout the year on a four to five day basis and obviously is subject to estrogenic influences to a much greater extent than dogs or cats, which cycle only on a seasonal basis. One study on dogs did reveal that spayed females gained about one kilogram (Houpt et al., 1979).

Male rats that are castrated undergo a moderate reduction in both food intake and body weight (Gentry and Wade, 1976). However, there is evidence that the weight loss is due to reduction of muscle mass, and that the percentage of body weight in fat actually increases (Mitchel and Keesey, 1974). Thus, our research information does modestly support the old notion that spaying females or castrating males tends to make them put on fat. The effect is not overwhelming. In fact, clinical surveys of the effects of castration on male dogs and cats found no indication that the owners felt the animals had gained weight (Hart and Barrett, 1973; Hopkins et al., 1976). Certainly adjusting the diet could counteract the rather modest effects of hormonal alteration.

Studies revealing that administration of estrogenic hormones decreases food intake and body weight in adult females contrast with the established procedure of the past of giving cattle and sheep the estrogenic chemical, diethylstilbestrol, in food to increase growth rate and meat quality (Baile and Forbes, 1974; Hafs et al., 1971). Apparently the weight-promoting effect is only on growing cattle and sheep, and is due to stimulation of pituitary growth hormone secretion rather than to an effect of estrogen on food metabolism.

■

ELIMINATIVE BEHAVIOR

Elimination, like feeding, is related to the diet and physiological characteristics of a species. Herbivores defecate very frequently, in the range of from 12 to 20 times per day for cattle and five to 12 times per day for horses. Carnivores defecate two to three times a day. There are two aspects of eliminative behavior that are of central importance. One is the location where animals urinate and defecate and the second is the social significance of urination and defecation.

For many wild mammals, as well as dogs and cats, elimination occurs not only as a necessary physiological process but also as a means of marking objects, such as trees, shrubs, and the exterior and interior of houses as an expression of territoriality (Ewer, 1968). The social or territorial aspect of elimination was discussed in Chapter 2.

Predisposition to Eliminate Away from the Sleeping Area

Some species, especially those having a home territory, have evolved behavioral tendencies to urinate and/or defecate away from the den, nest, or rest places. The natural selection for this behavior probably reflects a protection against intestinal parasites, which are transmitted among animals by eggs in the feces. With domestication, animal owners usually remove animal waste products, including parasite eggs, and medicate against intestinal parasites as well. Thus, one might argue that the reduced natural selection pressure allows the opportunity for a good deal of variability in sanitary behavior among domestic animals; some animals are fastidious in their attempts to keep a nest or bed clean whereas others may appear to be almost indifferent to where they eliminate.

There are several species-typical differences related to location for elimination. Swine, especially when confined in pens, have a strong tendency to eliminate away from the rest area. Often this is the part of the pen where the drinking trough is, which is already messed up. Selective dunging behavior of piglets occurs as early as three days of age. If the sow is able to move about freely, they eliminate in her dunging area. If she is restrained, as is usually the case on farms, they eliminate as far as possible from their sleeping areas (Whatson, 1978).

House dogs usually do not defecate or urinate in their sleeping and eating areas. Usually the entire house is treated as off-limits for elimination. Some dogs will defecate only in certain outdoor areas. Females do not tend to deposit urine in any particular location away from the home. Male dogs urinate on vertical objects in a scent-marking fashion, and when the tendency to mark is high, this may be done in the house.

Cats are difficult to classify in regard to location for elimination since both sexes tend to bury both urine and feces in loosely packed dirt or sand away from their nest. It is not uncommon to find a "well-broken" house cat that readily defecates in large planters or fireplace ashes as well as in a litter box.

Horses tend to interrupt activities such as eating and walking to urinate and defecate. In a field or paddock males and females usually defecate in one area, and when the pasture is large enough, this functions to keep most of the grazing area clean (Ödberg and Francis-Smith, 1977). However, in small fields there may be a problem as the dunging area enlarges. It is claimed that mares approach the dunging area and face it while defecating, thus progressively enlarging the area. Stallions back into the dunging area, thus restricting the area contaminated by feces. The dunging area of a pasture grazed only by mares may apparently get so large as to ruin it for grazing. Interestingly, it is claimed that castration of stallions results in a switch to the female type of behavioral pattern. In feral horses dunging areas function in communication.

Cattle, sheep, and goats do not consistently defecate or urinate with regard to location. In fact, these animals appear to have little voluntary control over the eliminative processes. This behavior is a reflection of the fact that the ancestors of domestic ruminants were not territorial and had little probability of returning to the same grazing area again in one season. Thus in such nonterritorial ruminants there is little chance of self-infestation with parasite eggs. Parasite control is accomplished by aversion to soiled foliage, as discussed previously. Males of territorial ruminants such as antelopes often have dunging areas for territorial marking.

There is a special posture used in defecation, especially in cattle, in which the tail is raised and deviated to the side, the back is raised in an arch, and the hind legs placed slightly forward. This posture reduces the soiling of the skin by feces. Sick cattle often defecate without showing this posture, and thus soiling of the skin may occur. In sheep and goats, the fecal pellets do not adhere to the skin and correspondingly no special posture is seen. Although males may urinate while walking, females tend not to. Males assume a posture involving lowering of the back, which prevents soiling of the skin by urine. Female goats and sheep assume a squatting posture when urinating.

Animals can learn to eliminate in certain areas or with certain postures. Cats, for example, have been taught to use a toilet by special conditioning procedures (Hart, 1975). Nonhuman primates have occasionally been taught to defecate in specific toilet receptacles, although normally they seem to have no tendency to eliminate in certain areas.

The electric barn trainers used to force dairy cattle to defecate in an alley behind the stanchion operate on the principle that the back must be arched for defecation to occur. By running an electric wire close above the back, the cows are taught to back up into the alley in order to defecate so as to avoid being shocked on the back.

One other aspect of eliminative behavior in dogs and cats is shown by the mother who keeps the nest area clean of feces and urine eliminated by the young animals. During the first few weeks of life the mother licks the anogenital region, evoking urination and defecation and consuming the material eliminated. Infant puppies and kittens are practically incapable of defecating and urinating without this type of stimulation; thus this behavioral system is very efficient in keeping the nest clean.

As the young animals grow older, but before they are capable of leaving the nest area, they are soon able to defecate and urinate without the licking stimulation provided by their mother. As long as they are unable to leave the nest the mother still consumes the material that is eliminated by the young, thereby continuing to keep the nest clean. As soon as the young are able to leave the nest area, defecation and urination usually occur away from the nest.

Female swine do not perform this function for the young in the nest. However, baby piglets are born quite mature compared with dogs and cats and are capable of moving about and eliminating away from the nest.

House Training Animals

Some of the major problems with eliminative behavior in pet dogs and cats involve house-training procedures. Theoretically, only animals that have an innate tendency to keep the nest or home area clean can be readily house-trained. This includes not only dogs and cats but other less domesticated species such as rabbits and skunks. Ruminants and birds are difficult but apparently not impossible to house-train if one employs special conditioning procedures. Even house training primates is extremely difficult, and successful instances seem to be the result of arduous systematic conditioning rather than the shaping of a natural tendency.

For cats house training merely consists of providing material of suitable texture such as sand or commercial litter, and the animal's native behavioral tendencies take it from there. Use of a sandbox may be learned more rapidly if the cat is restricted initially to one room with the box in a corner. Most cats will eliminate outdoors in sand or loose soil, if possible, when a litter pan is not placed inside. Cats do not seem to develop the same strong tendency to keep the house clean as a well-trained dog.

For dogs the house-training procedure can be thought of as a generalization of their natural tendency to defecate and urinate away from the nest or den area. A good house-training procedure should be successful with adult as well as with young dogs. One must realize that for a young animal the typical house may be much too large to be perceived as a "den" and that the animal is probably only initially capable of considering a corner of one room as a den. As the dog becomes more familiar with the house, it probably generalizes first to one or two rooms, then eventually to the entire house.

The most efficient house-training technique for puppies consists of restricting the animal to one small room or part of a room so that it gets the feeling that the small area is its nest. From this point it can be taken outdoors frequently, especially when the tendency to eliminate is high, such as after meals or sleeping. Under optimal circumstances the dog may never eliminate in the house.

References

ALBRIGHT, W. 1945. The story of four haystacks. *The Land* 4:228–232.

ARNOLD, G. W., AND M. L. DUDZINSKI. 1978. *Ethology of free-ranging domestic animals.* New York: Elsevier.

ARNOLD, G. W., AND R. A. MALLER. 1977. Effects of nutritional experience in early adult life on the performance and dietary habits of sheep. *Appl. Anim. Ethol.* 3:5–26.

BAILE, C. A., AND J. M. FORBES. 1974. Contrast of food intake and regulation of energy balance in ruminants. *Physiol. Rev.* 54:160–214.

BELL, F. 1959. Preference thresholds for taste discrimination in goats. *J. Agric. Sci.* 52:125–129.

BELL, F. 1960. The electroencephalogram of goats during somnolence and rumination. *Anim. Behav.* 8:39–42.

BELL, F., AND T. ITABISASHI. 1973. The electroencephalogram of sheep and goats with special reference to rumination. *Physiol. Behav.* 11:503–514.

BELL, F., AND A. LAWN. 1957. The pattern of rumination behaviour in housed goats. *Anim. Behav.* 5:85–89.

BERNSTEIN, I. L. 1978. Learned taste aversions in children receiving chemotherapy. *Science* 200:1302–1303.

BERNSTEIN, I. L., AND R. A. SIGMUNDI. 1980. Tumor anorexia: A learned food aversion? *Science* 209:416–417.

BOY, V., AND P. DUNCAN. 1979. Time-budgets of Camargue horses. Developmental changes in the time-budgets of foals. *Behaviour* 71:187–202.

BOYLE, P. C., L. H. STORLIEN, AND R. E. KEESEY. 1978. Increased efficiency of food utilization following weight loss. *Physiol. Behav.* 21:216–264.

BUENO, L., AND Y. RUCKEBUSCH. 1979. Ingestive behaviour in sheep under field conditions. *Appl. Anim. Ethol.* 5:179–187.

CAMPBELL, D., AND A. FRASER. 1961. A note on animal behaviour as a factor in parasitism. *Can. Vet. J.* 2:414–415.

CARR, W. J., J. T. HIRSCH, B. E. CAMPELLONE, AND E. MARASCO. 1979. Some determinants of a natural food aversion in Norway rats. *J. Comp. Physiol. Psychol.* 93:899–906.

COTE, P. J., AND P. J. WANGSNESS. 1978. Rate composition and efficiency of growth in lean and obese pigs. *J. Anim. Sci.* 47:441–447.

DELLA-FERA, M. A., AND C. A. BAILE. 1979. Cholecystokinin octapeptide: Continuous picomole injections into the cerebral ventricles of sheep suppresses feeding. *Science* 206:471–472.

DENTON, D. A. 1967. Salt appetite. In *Handbook of physiology*, vol. 1, sec. 6, pp. 433–459, ed. C. F. Code. Washington, D.C.: American Physiological Society.

DUDZINSKI, M. L., AND G. W. ARNOLD. 1979. Factors influencing the grazing behaviour of sheep in a Mediterranean climate. *Appl. Anim. Ethol.* 5:125–144.

DUDZINSKI, M. L., H. J. SCHUH, D. G. WILCOX, H. G. GARDINER, AND J. G. MORRISSEY. 1978. Statistical and probabilistic estimators of forage conditions from grazing behaviour of Merino sheep in a semi-arid environment. *Appl. Anim. Ethol.* 4:357–368.

ELLINS, S. R., S. M. CATALANO, AND S. A. SCHECHINGER. 1977. Conditioned taste aversion: A field application to coyote predation on sheep. *Behav. Biol.* 20:91–95.

EWBANK, R. 1969. Behavioural implications of intensive animal husbandry. *Outlook in Agric.* 6:41–46.

EWER, R. F. 1968. *Ethology of mammals.* New York: Plenum.

FONTENOT, J. P., AND R. E. BLASER. 1965. Symposium on factors influencing the voluntary intake on herbage by ruminants: Selection and intake by grazing animals. *J. Anim. Sci.* 24:1202–1208.

FORBES, J. M. 1971. Physiological changes affecting voluntary food intake in ruminants. *Proc. Nutr. Soc.* 30:135–142.

FOX, L. R. 1975. Cannibalism in natural populations. *Ann. Rev. Ecol. Syst.* 6:87–106.

FRIEDMAN, M., AND E. STRICKER. 1976. The physiological psychology of hunger: A physiological perspective. *Psychol. Rev.* 83:409–431.

GALEF, B. G., JR., AND M. M. CLARK. 1971. Social factors in the poison avoidance and feeding behavior of wild and domestic rat pups. *J. Comp. Physiol. Psychol.* 75:341–357.

GARCIA, J., W. G. HANKINS, AND K. W. RUSINIAK. 1974. Behavioral regulation of the milieu interne in man and rat. *Science* 185:824–831.

GENTRY, R., AND G. N. WADE. 1976. Androgenic control of food intake and body weight in male rats. *J. Comp. Physiol. Psychol.* 90:18–25.

GEYER, L. A., AND M. R. KARE. 1981. Taste and food selection in the weaning of nonprimate mammals. In *Food as an environmental factor in the genesis of human variability*, pp. 69–82, eds. D. Walcher and N. Kretchmer. New York: Masson Publishing Co.

GORDON, J. G., AND D. E. TRIBE. 1951. The self-selection of diet by pregnant ewes. *J. Agric. Sci.* 41:187–190.

GROSSMAN, S. P. 1975. Role of the hypothalamus in the regulation of food and water intake. *Psych. Rev.* 82:200–224.

GUSTAVSON, C. R., J. GARCIA, W. G. HANKINS, AND K. W. RUSINIAK. 1974. Coyote predation control by aversive conditioning. *Science* 184:581–583.

GUSTAVSON, C. R., D. J. KELLY, AND M. SWEENY. 1976. Prey-lithium aversions. I. Coyotes and wolves. *Behav. Biol.* 17:61–72.

HAENLEIN, G. F. W., R. D. HOLDREN, AND Y. M. YOON. 1966. Comparative response of horses and sheep to different physical forms of alfalfa hay. *J. Anim. Sci.* 25:740–743.

HAFEZ, E. S. E., AND M. F. BOUISSOU. 1975. The behaviour of cattle. In *The behaviour of domestic animals*, 3rd ed., pp. 203–245, ed. E. S. E. Hafez. Baltimore: Williams & Wilkins.

HAFS, H. D., R. W. PURCHAS, AND A. M. PEARSON. 1971. A review: Relationships of some hormones to growth and carcass quality of ruminants. *J. Anim. Sci.* 33:64–71.

HAMILTON, C. L. 1967. Food and temperature. In *Handbook of physiology*, vol. 1, sec. 6, pp. 303–317, ed. C. F. Code. Washington, D.C.: American Physiological Society.

HAMILTON, W. D. 1971. Geometry for the selfish herd. *J. Theoret. Biol.* 31:295–311.

HART, B. L. 1975. Learning ability in cats. *Feline Practice* 5(5):10–12.

HART, B. L. 1976. Quiz on feline behavior. *Feline Practice* 6(3):10–14.

HART, B. L., AND R. L. BARRETT. 1973. Effects of castration on fighting, roaming and urine spraying in adult male cats. *J. Amer. Vet. Med. Assn.* 163:290–292.

HENRY, J. D. 1977. The use of urine-marking in the scavenging behavior of the red fox (*Vulpes vulpes*). *Behaviour* 61:82–103.

HIRSCH, J., AND P. W. HAN. 1969. Cellularity of rat adipose tissue. Effects of growth, starvation, and obesity. *J. Lipid Res.* 10:77–82.

HIRSCH, J., AND J. L. KNITTLE. 1970. Cellularity of obese and nonobese human adipose tissue. *Fed. Proc.* 29:1516–1521.

HOPKINS, S. G., T. A. SCHUBERT, AND B. L. HART. 1976. Castration of adult male dogs: Effects of roaming, aggression, urine marking and mounting. *J. Amer. Vet. Med. Assoc.* 168:1108–1110.

HOUPT, K. A., B. COREN, H. F. HINTZ, AND J. E. HILDERBRANT. 1979. Effect of sex and reproductive status on sucrose preference, food intake, and body weight of dogs. *J. Amer. Vet. Med. Assn.* 174:1083–1085.

JACOBS, W. W., AND J. N. LABOWS. 1979. Conditioned aversion, bitter taste and the avoidance of natural toxicants in wild guinea pigs. *Physiol. Behav.* 22:173–178.

JESELNIK, D. L., AND I. L. BRISBIN. 1980. Food-caching behaviour of captive-reared red foxes. *Appl. Anim. Ethol.* 6:363–367.

KUO, Z.-Y. 1967. *The dynamics of behavioral development.* New York: Random House.

LANDAU, I. T., AND I. ZUCKER. 1976. Estrogenic regulation of body weight in the female rat. *Horm. Behav.* 7:29–39.

LESHNER, A., AND G. COLLIER. 1973. The effects of gonadectomy on the sex differences in dietary self-selection patterns and carcass compositions of rats. *Physiol. Behav.* 11:671–676.

LESHNER, A. I., G. H. COLLIER, AND R. L. SQUIBB. 1971. Dietary self-selection at cold temperatures. *Physiol. Behav.* 6:1–3.

LOBATO, J. F. P., G. R. PEARCE, AND R. G. BEILHARZ. 1980. Effect of early familiarization with dietary supplements on the subsequent ingestion of molasses-urea blocks by sheep. *Appl. Anim. Ethol.* 6:149–161.

LOFGREN, G. P., J. H. MEYER, AND J. L. HULL. 1957. Behavior patterns of sheep and cattle being fed pasture on soilage. *J. Anim. Sci.* 16:773–780.

LOTT, A. F., AND B. L. HART. 1979. Applied ethology in a nomadic cattle culture. *Appl. Anim. Ethol.* 5:309–319.

MARGULES, D. L., B. MOISSET, M. J. LEWIS, H. SHIBUYA, AND C. B. PERT. 1978. β-endorphin is associated with overeating in genetically obese mice (*ob/ob*) and rat (*fa/fa*). *Science* 202:988–991.

MITCHEL, J. S., AND R. E. KEESEY. 1974. The effects of lateral hypothalamic lesions and castration upon the body weight and composition of male rats. *Behav. Biol.* 11:69–82.

MONTGOMERY, G. W., D. S. FLUX, AND J. R. CARR. 1978. Feeding patterns in pigs: The effects of amino acid deficiency. *Physiol. Behav.* 20:693–698.

MORLEY, J. E., AND A. S. LEVINE. 1980. Stress-induced eating is mediated through endogenous opiates. *Science* 209:1259–1261.

MROSOVSKY, N., AND D. F. SHERRY. 1980. Animal anorexias. *Science* 207:837–842.

MUGFORD, R. A. 1977. External influences on the feeding of carnivores. In *The chemical senses and nutrition*, eds. M. R. Kare and O. Maller. New York: Academic.

MULLER, L. D., L. V. SCHAFFER, L. C. HAM, AND M. J. OWENS. 1977. Cafeteria style free-choice mineral feeder for lactating dairy cows. *J. Dairy Sci.* 60:1574–1582.

NACHMAN, M. 1962. Taste preferences for sodium salts by adrenalectomized rats. *J. Comp. Physiol. Psychol.* 55:1124–1129.

NISBETT, R. 1972. Hunger, obesity, and the ventromedial hypothalamus. *Psychol. Rev.* 79:433–453.

ÖDBERG, F. O. 1975. Eliminative and grazing behaviour in the horse. *Appl. Anim. Ethol.* 1:203–209.

ÖDBERG, F. O., AND K. FRANCIS-SMITH. 1977. Studies on the formation of ungrazed eliminative areas in fields used by horses. *Appl. Anim. Ethol.* 3:27–34.

OVERMANN, S. R. 1976. Dietary self-selection by animals. *Psychol. Bull.* 83:218–235.

PANSKEPP, J. 1971. Effects of fats, proteins, and carbohydrates on food intake in rats. *Psychonom. Mono. Suppl.* 4:85–95.

PYKE, G. 1978. Are animals efficient harvesters? *Anim. Behav.* 26:241–250.

RICHTER, C. P. 1943. Total self-regulatory function in animals and human beings. *Harvey Lectures Series* 38:63–103.

RILEY, A., AND D. ZELLNER. 1978. Methylphenidate-induced conditioned taste aversions: An index to toxicity. *Physiol. Psychol.* 6:354–358.

ROGERS, Q. R., AND A. E. HARPER. 1970. Selection of a solution containing histidine by rats fed a histidine-imbalanced diet. *J. Comp. Physiol. Psychol.* 71:66–71.

ROGERS, Q. R., AND P. M. B. LEUNG. 1977. The control of food intake: When and how are amino acids involved? In *The chemical senses and nutrition*, eds. M. R. Kare and O. Maller. New York: Academic.

ROSS, S., AND J. ROSS. 1949a. Social facilitation of feeding behavior in dogs. I. Group and solitary feeding. *J. Genet. Psychol.* 74:97–108.

ROSS, S., AND J. ROSS. 1949b. Social facilitation of feeding behavior in dogs. II. Feeding after satiation. *J. Genet. Psychol.* 74:293–304.

ROZIN, P., AND J. KALAT. 1971. Specific hungers and poison avoidance as adaptive specializations of learning. *Psychol. Rev.* 78:459–486.

ROZIN, P., AND W. RODGERS. 1967. Novel diet preferences in vitamin deficient rats and rats recovered from vitamin deficiencies. *J. Comp. Physiol. Psychol.* 63:421–428.

RUCKEBUSCH, Y., AND L. BUENO. 1978. An analysis of ingestive behavior and activity of cattle under field conditions. *Appl. Anim. Ethol.* 4:301–313.

SEOANE, J., C. A. BAILE, AND F. H. MARTIN. 1972. Humoral factors modifying feeding behavior of sheep. *Physiol. Behav.* 8:993–995.

SIGNORET, J. P., B. A. BALDWIN, D. FRASER, AND E. S. E. HAFEZ. 1975. The behaviour of swine. In *The behaviour of domestic animals*, 3rd ed., pp. 295–329, ed. E. S. E. Hafez. Baltimore: Williams & Wilkins.

SQUIRES, V. 1971. Temporal patterns of activity in a small flock of Merino sheep as determined by an automatic recording technique. *Anim. Behav.* 19:657–660.

TARTTELIN, M. R. 1968. Cyclical variations in food and water intake of ewes. *J. Physiol.* 195:29P–31P.

TINBERGEN, N. 1972. Food hoarding by foxes (*Vulpes vulpes* L.). In *The animal and its world*, vol. 1, *Field studies*, pp. 315–318. Cambridge: Harvard University Press.

TRIBE, D. E. 1955. The behaviour of grazing animals. In *Progress in the physiology of farm animals*, p. 585, ed. J. Hammond. London: Buttersworth.

WADE, G. N. 1976. Sex Hormones, regulatory behaviors, and body weight. In *Advances in the study of behavior*, vol. 6, pp. 201–279. New York: Academic.

WANGSNESS, P. J., J. L. GOBBLE, AND G. W. SHERRITT. 1980. Feeding behavior of lean and obese pigs. *Physiol. Behav.* 24:407–410.

WHATSON, T. S. 1978. The development of dunging preferences in piglets. *Appl. Anim. Ethol.* 4:293.

WIER, W. C., AND D. T. TORELL. 1959. Selective grazing by sheep as shown by a comparison of the chemical composition of range and pasture forage obtained by hand clipping and that collected by esophageal fistulated sheep. *J. Anim. Sci.* 18:641–649.

WILLARD, J., J. C. WILLARD, AND J. P. BAKER. 1973. Dietary influence on feeding behavior in ponies. *J. Anim. Sci.* 37:277.

WILLARD, J. G., J. C. WILLARD, S. A. WOLFRAM, AND J. P. PARKER. 1977. Effect of diet on cecal pH and feeding behavior of horses. *J. Anim. Sci.* 45:87–93.

Genetics is an integral part of the consideration of the behavior of domestic animals. In fact, domestication implies genetic control over behavior. The difference between a tame wild animal and a domesticated one is that in the latter, changes such as reduction of fear and aggressive reactions toward people have been accomplished by selective breeding. Thus, the species-typical characteristics for the domesticated breed are actually modified. In contrast, taming implies that behavioral changes are brought about only within an animal's lifetime by early experiences, learning, and the physical constraints of the domestic environment. In this chapter we shall examine the ways in which genetic manipulations can have major influences on behavior. This chapter, on genetic control of behavior, and Chapters 7 and 8 on early experience and learning illustrate the range of influences that may affect behavior of domestic animals.

The influence of genetics on behavioral patterns in domestic animals should receive more attention than it does. In animals kept for food and fiber and in companion animals, there are frequently undesirable behavioral traits that might well be altered or avoided by attention to genetic influences. Throughout Chapters 2, 3, 4, and 5 on species-typical behavior patterns, we have seen, repeatedly, changes in behavior that presumably reflect relaxation of natural selection pressures. In addition to preventing some undesirable changes, there is also the possibility of enhancing some behavioral characteristics to improve an animal's suitability to the domestic environment.

The three major species of companion animals—horses, cats, and dogs—are kept because people are attracted to their behavior. Yet in breeding these animals, behavior receives considerably less attention than morphological characteristics. This is not to say there are not major behavioral differences among breeds; there most definitely are. But breeds are generally defined on the basis of body size and shape or hair texture and color. Available evidence indicates that behavioral traits can be influenced as easily as anatomical or physiological traits. Thus conceivably a breed of cats, dogs, or horses could be produced that would be primarily characterized by behavior rather than morphology. A special breed of dogs could be developed for urban environments that might have a high threshold of barking, low emotional reactivity, and a very strong tendency toward not defecating or urinating in the home. The ideal breed for a farm dog would be developed along different lines. The urban breed of cat might be one with a very low probability of catching songbirds. Horses might be bred with a particular emphasis on gentleness and ease of handling.

Work with mice and rats has revealed that behavioral characteristics may be carried by animals even though they are not displayed under natural conditions. Alcohol preference in some mouse strains is a trait that has certainly not been subjected to the influences of natural selection, yet there are clear differences among inbred strains (McClearn and Rogers, 1959, 1961). A similar demonstration of the same concept was revealed in an experiment showing selective breeding for saccharin preference (Nachman, 1959).

Millions of domestic animals are kept under a variety of household and farm conditions. It would appear as though there is ample opportunity for the occasional occurrence of an unusual but highly desirable behavioral pattern. Selective breeding could capitalize on such behavior and possibly set the stage for its general dissemination. Such traits might have to do with maternal behavior in livestock species, hunting behavior in dogs, or temperament in horses.

The purpose of this chapter is not to present a detailed discourse on breeding and behavioral genetics, but to view the principles of genetics from the standpoint of domestic animal behavior.

■

PRINCIPLES OF BEHAVIORAL GENETICS

Although some of the same principles that apply to genetic control of morphological and physiological characteristics can be applied to behavioral responses, behavioral genetics is not merely a subcategory of genetics. Behavior has been studied very intensely and in a manner different than the study of coloration, body morphology, or milk production. In other areas of genetics, one might be primarily interested in the mechanisms of inheritance, such as linkage or the number of genes involved. This is not the primary focus in behavioral genetics (Thompson, 1967). The critical genetic considerations are those that shed some light on understanding behavioral patterns. Behavioral genetics also differs from other areas of genetic research because behavior itself is complex. Practically any behavioral characteristic having a genetic basis is also going to be influenced by environmental factors, which include learning and early experience. And the two influences—genetics and environment—must be considered hand in hand because genetic influences interact with those from the environment. In many instances one cannot tell just by looking at a behavior whether it is primarily the result of inheritance or of learning. Furthermore, most behavioral characteristics are influenced by several genes (polygenic traits).

One principle that is sometimes overlooked is that behavioral genetics is not concerned with the inheritance of behavioral traits per se, but rather with behavioral differences between animals or groups (strains or breeds) of animals. We are not concerned, for example, with whether animals inherit aggressive tendencies, but rather whether a cluster of behavioral patterns we call aggressive differs between breeds, and whether this difference can be attributed to genetic or environmental influences. Behavioral genetics concentrates on determining the types of behavior that are influenced by inheritance, and the degree to which behavioral differences between individuals are due to genetic as compared with environmental factors. The richness of genetic effects on behavior is due to the great degree of individual variability in the genetic makeup of animals.

■

PHYSIOLOGICAL MECHANISMS OF GENETIC INFLUENCES ON BEHAVIOR

A few comments should be made about the physiological link between the gene and behavior. Behavior itself, of course, is not inherited. Behavior is mediated by the central nervous system, which in turn has developed and differentiated under the control of genes. For all of the individuals in a species the general pattern of nervous system development is the same. But like any biological system, there is variation in the brain between individuals in a species with regard to anatomical, physiological, and biochemical details. This is presumably what results in individual differences in behavior. There is some reason to believe that the same anatomical, physiological, and biochemical differences may result from either genetic or environmental influences. That is, the influences of early experience, learning, or hormones may produce some of the kinds of changes in an individual's brain that would be produced by selective breeding. Thus, even at the molecular level, there is not necessarily a qualitative difference between genetic and environmental influences on behavior.

Some attempts have been made to identify particular biochemical or anatomical correlates of genetic influences over behavior. For example, the biochemical link between a gene and a condition in humans known as phenylketonuria, which is characterized by mental retardation, is known to involve a deficiency of the enzyme phenylalanine hydroxylase. In another line of research it has been found that genetically controlled audiogenic seizure susceptibility in mice is related to an enzymatic effect on nucleoside triphosphatase in the granular layer of the dentate fascia of the hippocampus (Ginsburg, 1967).

Types of Effects of Heredity on Behavior

Investigators have been concerned with determining what types of behavioral characteristics can be most easily influenced in animals. Sexual, aggressive, exploratory, emotional, and maternal behavior as well as such diverse patterns as problem-solving ability and alcohol preference have received the attention of behavioral geneticists. Practically any behavioral trait that can be quantitatively measured, and that has been systematically studied, has been shown to be influenced by genetic factors.

To demonstrate what particular behavioral characteristics are indeed influenced by genetic elements, investigators have used two approaches. In one approach, animals of mixed or heterogeneous genetic composition are measured for a behavioral characteristic and then the individuals that perform at both extremes of the behavior are selectively bred and their progeny measured for the same behavior. After a number of generations resulting from such artificial selective breedings, quantitative differences may be shown. It is assumed that this is proof of genetic control. The second approach is to use already established inbred or relatively homogeneous strains and simply look for behavioral differences as a function of strain identification.

Selective Breeding   One of the better-known examples of selective breeding of heterogeneous animals for a particular behavioral trait is Tryon's experiment in breeding "maze-bright" and "maze-dull" rats. The divergence in behavioral performance as a function of selection after each generation is illustrated in Figure 6-1.

One valuable lesson in behavioral genetics was revealed in a further comparison of Tryon's two strains (McClearn, 1963). One might have a tendency to think of maze learning as an indication of intelligence or at least of learning ability, and of the selective breeding as producing strains differing in intelligence. This was not the case; it was found that the maze-bright animals were no better at other, nonmaze types of learning tasks. Several other comparisons of the maze-bright and maze-dull strains have revealed little evidence of a direct genetic influence over general learning ability (Wahlsten, 1972).

By selective breeding, investigators have established strains of rats or mice differing in speed of avoidance learning, general activity, emotionality, and aggressive behavior.

Some work on the genetic influence on various components of sexual behavior is particularly interesting. The genetic factors influencing the quantitative aspects of various components of sexual behavior are relatively independent (Goy and Jakway,

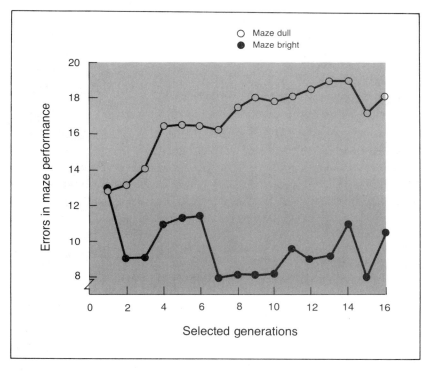

Figure 6-1 A classic selective breeding experiment by Tryon showed that rats could be bred for "maze-bright" or "maze-dull" behavior. The graph presented here shows scores in maze performance with each succeeding generation. (From McClearn, 1963.)

1962). For example, mounting activity may be altered independent of ejaculation latency. The genetic separation of components is true for both male and female sexual behavior, and some aspects of male sexual behavior may be genetically manipulated independently of any influences on female sexual behavior. Many authorities feel that the demonstration of the potential to produce behavioral characteristics to specification in experimental animals indicates that practically any trait can be enhanced by selection (McClearn, 1967).

Analysis of Inbred Strains   The second approach of identifying what behavioral patterns can be influenced involves the analysis of existing inbred strains for differences in behavioral traits. This work has been most extensive in mice, in which some strains have virtually complete homozygosity. It has been shown that

C3H mice are more likely to initiate aggression than C57BL mice (Scott, 1942). However, C57BL mice are superior to C3H animals in their ability to win fights (Ginsburg and Allee, 1942). Analysis of sexual activity in mice, especially in frequency of ejaculation per test, has revealed that C57BL mice are superior to animals of the DBA strain (McGill and Tucker, 1964).

Other types of strain differences that have been reported are differences in susceptibility in rats to ulceration in a conflict situation and differences in general activity, learning, and emotional responses in various strains of mice (Broadhurst and Levine, 1963; Bruell, 1967; Thompson, 1953). In one interesting study, it was learned that the effect of stress administered to pregnant female mice caused a significant change in the general activity of the offspring. However, in one strain the result was a reduction in activity and in another strain there was an increase in activity (Weir and DeFries, 1964).

Measurement of the Effects of Heredity on Behavior

There are different degrees to which various traits (phenotype) are influenced by genetic make-up (genotype). Hair color and blood type are examples of almost complete determination of a trait by genetic elements. Body size is an example of a characteristic that is influenced partially by genetics and partially by environmental influences such as nutrition, exercise, and diseases.

When a particular behavioral trait is influenced by a number of genes, the genetic effect is referred to as additive in those instances where the various combinations of genes may summate. However, there is evidence that different combinations of genes may result in a different type of genetic-environmental interaction, and the additive model may be an oversimplification. It is possible for the same behavioral phenotype to be produced by several different genotypes, just as the same behavioral phenotype may be produced by a number of different environmentally oriented approaches.

Genetic influences over biological traits in general may also be expressed in a discontinuous way as in the inheritance of blood type or in a continuous way as in the inheritance of human skin pigmentation. Most behavioral characteristics are controlled by multiple genes and are expressed in a continuous rather than a discontinuous way. Within the framework of continuous genetic control, most behavioral traits are expected to show intermediate inheritance. By this is meant that quantitative measures of the behavior of the progeny are intermediate between measures of the same behavior in the parents, just as skin pigmentation in people generally falls in between that of the parents. Bringing environment back into the picture again, behavioral characteristics are a reflection of intermediate inheritance of characters on a continuum and of the environment, just as human skin color is expressed as intermediate inheritance and is affected by sunlight.

Behavioral Heterosis   An interesting aspect of behavioral genetics involves the occurrence of a behavioral phenotype in the offspring that exceeds that of either parent. To the extent that a phenotype is influenced by environmental factors, the superior performance could be explained by learning or nutrition. However, in several instances where environmental conditions have been controlled, quantitative measures of some behavioral traits have exceeded those of inbred parents. This phenomenon is known as hybrid vigor, or heterosis. It can be contrasted with the intermediate inheritance of behavior, where the offspring's performance falls between that of the parents.

A number of examples of behavioral heterosis have been described. Caged mice belonging to different strains differ in the amount of running in exercise wheels they do. The hybrid offspring from mating between active strains and inactive strains generally exhibit heterosis by outrunning both parents. As is illustrated in Figure 6-2, most hybrid offspring of various inbred crosses achieved scores exceeding those of the parents (Bruell, 1967). Exploratory behavior of mice as tested in a four-compartment exploration maze also appears to be heterotic. Heterotic inheritance of avoidance conditioning and success in food competition in mice have also been reported (Collins, 1964; Manosevitz, 1972).

An example of a very pronounced effect of behavioral heterosis in the persistence of sexual behavior after castration was revealed in a study by McGill and Tucker (1964). In tests of mating behavior of two inbred strains of mice, they found that copulatory behavior disappeared very soon following castration. In one strain, the last ejaculation occurred an average of three days after castration; animals in the other strain did not copulate at all. In animals of the hybrid cross heterosis was displayed by

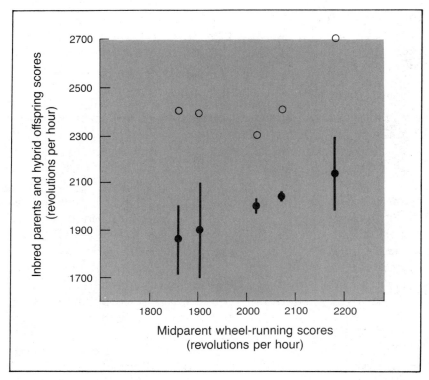

Figure 6-2 This graph of wheel running of inbred mice and their hybrid offspring shows the effect of heterosis. Scores of inbred parents are at the ends of the vertical bars, with a solid circle marking the midparent score. Scores of hybrid offspring of inbred crosses are shown directly above parent scores. (From Bruell, 1967.)

the persistence of copulatory behavior (ejaculation) an average of 28 days after castration.

Not all behavioral traits show heterosis, and it is difficult to predict which traits will. Presumably, heterosis is usually seen under conditions of crossing lines of unrelated inbred strains for those traits such as learning, food competition, sexual behavior, and general activity (wheel running), that enhance fitness or adaptation of the environment (Bruell, 1967).

Heritability: Measuring Genetic Influences   Studies and observations of behavior were at one time designed to determine if a particular pattern was controlled by inheritance (nature) or environment (nurture). Although such simplistic approaches are criticized now, at least they represented a first-order approach to an understanding of the multifaceted determinants of behavior. One of the most useful current concepts that takes into account the variety of genotypes and the interactions of these genotypes with the environment is that of heritability.

The concept of heritability comes back to the point made previously that behavioral genetics involves an analysis of differences in behavior between animals rather than the behavior per se of a single animal. Heritability is an expression that represents the degree to which variation in behavior can be attributed to variation in the genetic makeup of individuals. In a population in which the genetic makeup of all individuals is identical, all the variation has to have been produced by the environment. On the other hand, if the environment could be held constant, all the variation in a group would be due to genetic variability. Since it is impossible to achieve complete environmental consistency, and it is usually not possible to study animals that have complete genetic consistency, estimates of genetic variability must be derived indirectly.

The mathematical concept of heritability is based on the statistical measurement of variance as a specification of variation within a population. Variance is the square of the standard deviation, which is the commonest way of expressing dispersion about a mean (the method of calculating standard deviation is available in a number of books dealing with statistical analysis). Measures of a trait in two groups of individuals may reveal the same mean but have very different standard deviations.

The following is a simple expression of heritability:

$$\text{Heritability} = \frac{V_g}{V_p} \quad \text{or} \quad \frac{V_g}{V_g + V_e + V_i}$$

where

V_g = Variance in the population attributable to genetic factors

V_e = Variance in the population attributable to environmental factors

V_i = Variance in the population attributable to the interaction of genetic with environmental factors

V_p = Total variance of the population

This equation merely says that heritability, which may range from 0 to 1.0, is the result of dividing the variance attributable to genetic factors by the total variance of the population. The population variance is the sum of the variances attributable to genetic and environmental factors plus whatever effects may result from the interaction of these two influences. For general purposes it may be assumed that genetic and environmental influences are relatively independent and additive (noninteracting) and the V_i term can therefore be neglected. A heritability of 0 means that all the variability within the population is due to environmental factors, and a heritability of 1 means that all of the population variability is due to genetic factors.

For the purpose of explaining the concept one might consider a hypothetical example of grooming frequency of infants by wild female mice. Let us say that the mean number of daily grooming episodes is 64, and the variance is 25 (standard deviation of 5). If we now turn to a certain inbred strain, imagine that the mean number of daily grooming episodes is 58 with a variance of 16 (standard deviation of 4). Since the inbred strain is genetically homogeneous, all of the variance in the inbred strain must be due to environmental factors. The V_e for the wild group can be estimated by substituting the V_e of the inbred strain. Thus for the wild population

$$V_e = 16 \text{ (from testing the inbred strain)}$$

$$V_p \text{ or } V_g + V_e = 25 \text{ (from testing the wild population)}$$

and

$$V_g = 25 - 16 = 9$$

$$\text{Heritability} = \frac{V_g}{V_p} = \frac{9}{25} = 0.36$$

In domestic animals homogeneous inbred strains do not exist, and heritability is customarily determined by obtaining estimates of V_g and V_e from correlations of behavioral measures from relatives. Members of kinships have more genes in common than members of the population at large. Estimates of heritability are derived from parent, offspring, and sibling comparisons with the assumption that environmental deviations of the relatives are not correlated. This is not necessarily true, since relatives such as littermates have a similar environment and share similar climatic, nutritional, and experiential conditions. Rather sophisticated techniques for the estimation of heritability have been discussed at length by McClearn and DeFries (1973), Roberts (1967), and Falconer (1960).

An example of a measure of heritability using sibling correlation is the heritability of avoidance conditioning in swine. In an experiment the behavioral task required young pigs to jump from one side of a wooden box to another after the onset of a buzzer in order to avoid being shocked six seconds after the buzzer. There was considerable variance in the avoidance responses of the pigs, and the investigators obtained a heritability measure of about 0.45 (Willham et al., 1963). Estimates of heritability from similar experimental tests of avoidance behavior in mice and rats have also yielded heritability measures of about 0.5 (Wahlsten, 1972).

A retrospective procedure for estimating heritability that can be used along with a breeding program involves comparing actual improvement in a behavioral task with that which would be expected if heritability were 1. To illustrate the procedure, consider the theoretical possibility of selectively breeding cats for a low tendency to capture songbirds. Cats could be placed for an hour or two once a day in a large enclosure containing a number of birds. It might then be feasible to count the number of birds captured each week by various individuals and then selectively breed those individuals that kill the smallest number of birds. Heritability could then be calculated on the basis of comparing the realized reduction in the mean number of birds captured by the progeny with number of birds captured by the parents.

A more effective and humane approach to breeding cats for a low tendency to catch birds would be to devise a scoring system to reflect only elements of bird-catching behavior and to determine a mean daily score for each animal. For example, the elapsed time before a cat oriented toward a bird could be measured.

Cats could then be placed in the bird enclosure for daily 15-minute periods for scoring. Say that for a colony of 20 cats the mean score, after repeated tests on each animal, was two minutes, with a range of from one to 15 minutes. The five females and one male with the highest scores might then be selected as the breeding group. The progeny could then be retested as adults. Assume that the mean score of the breeding group was seven minutes, that is, five minutes above the mean for the initial colony. With a heritability of 1, the progeny would also have a mean score of seven minutes or a gain of five minutes over the initial colony. If in an actual breeding trial the progeny had a mean score of five minutes, then the gain over the original population would be three minutes. The realized heritability would be 3/5 or 0.60.

A more difficult determination of heritability arises when a behavior can only be observed in one sex. Consider the behavior problem in swine of sows that are likely to lie on one or more piglets and smother them (see Chapter 4). It appears as though some sows look before they lie down and hence avoid smothering piglets, whereas others lie down with a total lack of concern. One might be interested in breeding a strain of pigs with the tendency to look before lying down. It would be necessary to set up a maternal behavior scoring system reflecting a sow's attentiveness to baby piglets. One could record, for example, the percentage of times a sow carefully examines her litter before lying down. A score of over 90 percent might reflect the most desirable maternal behavior and a score below 10 percent the least desirable. Assume the mean for a group of sows was 35 percent with a range of 20 percent to 95 percent. One might select for a breeding group several sows in which the mean score was 85 percent. This behavior cannot, of course, be measured in males, even though the male probably contributes genetic influences to this aspect of maternal behavior in females. Males could be chosen for their genetic contribution to maternal care through testing the maternal behavior of their female offspring (progeny testing), but assuming this is not done, any improvement must be attributed to females. Because we are relying on only females to pass on good mothering genes, the heritability must be divided in half.

An actual study involving the determination of realized heritability was conducted on aggressive behavior in chickens by Guhl et al. (1960). By breeding and selecting chickens for aggressiveness on the basis of fights won and social rank, substantial improvements were made by the second generation (Figure 6-3). The mean heritability after four generations, based on selection for fight encounters won, was 0.22. The same type of improvement was found in selection of aggressiveness with White Leghorn and Rhode Island Red chickens with realized heritabilities of 0.16 and 0.28 respectively (Craig et al., 1965).

Heritability is a very useful concept in animal breeding. The higher the heritability estimate, the more one can expect to achieve significant results by selective breeding. Even heritabilities as low as 0.20 may be enough to allow for significant alteration of a behavioral pattern. Some of the mathematical

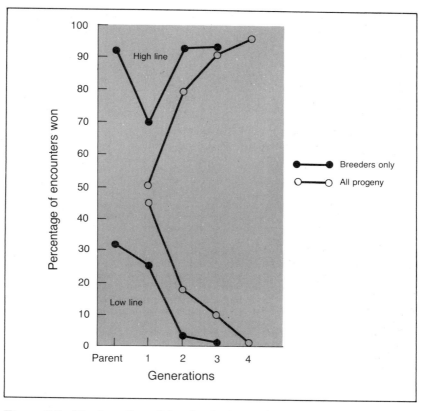

Figure 6-3 The breeding of female chickens for high and low ability to win fight encounters is shown. The percentage of encounters won by progeny selected as breeding stock is indicated by the line with solid circles, and the percentage of encounters won by all progeny is indicated by the line with open circles. (From Guhl et al., 1960.)

and behavioral principles involved in the determination of realized heritability have been reviewed by Fuller (1962). He emphasizes that a single determination of realized heritability is subject to error because of differing environmental conditions. Thus, in experimental work controls should be employed, and heritability determination should be carried out over several generations.

As the examples here point out, the application of genetic principles to behavior is not simply a matter of selective breeding. Procedures for measuring behavioral responses and for collecting reliable data with which to make comparisons across several generations are essential. Behavioral traits with relatively high phenotypic variances, but with very low heritabilities, will respond more to environmental than genetic manipulation. For breeding purposes, one should compare the heritability of a behavioral trait with that of morphological and physiological traits that have been successfully influenced by breeding in the past to get an idea of the practical feasibility of selective breeding for behavioral results. It must be understood that the estimate of heritability applies to one population in a specific environment and that the heritability measure cannot necessarily be extrapolated to another population. However, the more similar genetic and environmental characteristics are, the more similar the estimates of heritability will be.

■

GENETIC CONSIDERATIONS IN ANIMAL DOMESTICATION AND BREEDING

Domestic animals appear to possess few behavioral patterns that are not found in the wild ancestors. But domestication does imply genetic control over behavioral as well as physiological and morphological characteristics. By domestication we have accentuated some traits and reduced or eliminated others rather than create completely new traits. Thus, domestication has affected behavior quantitatively rather than qualitatively. The various ramifications of the effects of domestication on animal behavior, including behavioral changes, effects of environmental versus genetic manipulation, and feralization (the reverse of the domestication process) are treated in an excellent review by Price (1984).

Behavioral Aspects of Domestication

Under natural conditions, those behavioral traits were selected for that resulted in the greatest potential for escaping from predators, obtaining food, or breeding with the largest number of females, thereby favoring animals who produced the most offspring or maximized their inclusive fitness. In domestic animals, selection for behavioral traits is guided by adaptation to the farm, home, or laboratory environment. The animals that are the most desirable under these conditions are the ones that are used in breeding. Thus domestic animals, including livestock species, companion animals, and the laboratory rat, are generally more docile and more easily handled and restrained than their wild counterparts. One of the reasons exotic animals make such poor pets is that although they might be tamed, their natural fear of people and confinement and their flight or fight tendencies still exist.

Up until the widespread use of artificial vaginas to collect semen from males and artificial insemination to breed females, it was claimed that males of domestic species were sexually more vigorous than their wild ancestors because the most sexually vigorous males were used for farm breeding. Nowadays we can almost bypass natural breeding, and we tend to select males and females only for economic production as measured in weight gain, milk yield, litter size, and so forth irrespective of the animal's sexual prowess.

It has been asserted that juvenile behavioral characteristics in dogs and cats such as the tendencies to engage in play behavior and to seek affection appealed to the people who were initially involved in domesticating animals. Hence breeding practices tended to enhance this characteristic. The process of accentuating juvenile behavior in adult animals through selective breeding is referred to as neoteny. This trait is, of course, more evident in some breeds of dogs than others (Frank and Frank, 1982).

The other side of the coin regarding genetic changes produced by domestic husbandry practices is the frequently noted deterioration in performance of innate behaviors that were essential for survival in the wild. We have seen many instances in earlier chapters of impairments in behavior stemming from the fact that we step in and make up for such behavioral deficiencies.

When we medically treat for intestinal parasites, fastidiousness in keeping the nest clean is no longer a determinant of survival rate or reproductive success. The same goes for deficiencies in maternal behavior when we hand-raise neglected puppies and kittens or build farrowing crates to keep unattentive sows from lying on their piglets.

When a behavioral pattern such as sexual performance, maternal responsiveness, or diet selection is essential to survival or reproductive success, all individuals possess the behavior, and the behavior shows very little variability among individuals. In fact, because of the lack of variability, heritability of such traits is almost zero. We tend to call such traits instinct, implying that all animals display the behavior. When the natural selection pressures are removed, when display of the behavior becomes almost irrelevant to an animal's survival or reproductive success, variability in performance of the behavior creeps in because the deviants harboring genetic mutations, chromosome crossovers, and so forth do not meet a dead end. The resulting variability in these traits means there is a higher heritability than one would find in the wild ancestors.

The picture that has just been painted is one of behavioral degeneracy. As one might expect, the laboratory rat has been used as the subject for a fuller examination of the issue, but the concept can apply to most domestic species. The laboratory rat is, in fact, a highly domesticated version of the wild Norway rat. In a classic paper, Lockard (1968) has this to say about the laboratory rat:

> The phylogenetically established behaviors essential to survival in the wild are no longer kept intact in peak condition. Many, such as attack and flight from big species like man, intraspecific aggression, and strong pair bonds, are maladaptive in the new environment and are vigorously selected against. Phenotypes with these traits leave few progeny, and the genes responsible for the behavior decline in frequency and may vanish from the population. A great number of other genetically determined properties are subject to either low or neutral selection pressure; they are not disadvantageous, just unnecessary. There is little apparent reproductive advantage to keen sensory function, predator recognition, dietary preferences, digestive efficiency, heat-conserving nests, complicated burrows, or accurate behavioral periodicities. These sorts of functions, or characters, would more or less persist, but with greatly increased variability. Mutations affecting them would be preserved and accumulate in the gene pool. The net results would fall along a spectrum from complete loss of the function through degenerate, incomplete, or inappropriate expression, to apparent ineptness, depending upon the specific case. On the other hand, the new environment is not merely a route to tame degeneracy; it makes various positive demands, vigorously selecting for genes favoring survival and differential reproduction in captivity, the new 'natural habitat.' Examples of the sorts of traits favored might be intraspecific 'amiability,' immediate acceptance of strangers, large litters, long reproductive period, year-round fertility, reduced mobility, tolerance of loud sounds and handling by humans.

The notion that domestication of the laboratory rat or livestock and pets represents a type of degeneracy somewhat like that of weakened muscles or obesity is debatable. For example, there are a number of tasks at which domestic rats perform equal to or better than wild rats. They can dominate wild rats, perform better in some learning tasks, and show more behavioral plasticity (Boice, 1972, 1973; Boreman and Price, 1972; Mason and Price, 1973; Price, 1972). Boice (1973) states the opposite viewpoint quite effectively: "It is illogical to judge the domestic animal as unfit or degenerate because he is not adapted to live in a wild environment. The equivalent would be to label undomesticated animals as degenerate because they are not adapted for the 'superior' environment created by man."

In fairness to both sides of the question, what should probably be recognized is that so-called deficiencies in behaviors do not represent irreversible degeneracy but ongoing reactions to our breeding. If we find certain changes undesirable, we should give more attention to those traits in breeding programs.

Behavioral Aspects of Breed Identification

One of the obvious results of domestication is the establishment and maintenance of certain animal breeds. Typically these breeds have been established along morphological lines such as body conformation or coloration. Physiological characteristics have also served as a basis for breed identification; for example, the butterfat content of milk in cows or racing performance in horses. Although there is a good deal of genetic uniformity among breeds of domestic animals, they are not as homogeneous as strains of laboratory mice or rats. In the previous discussion of inbred strains of mice and rats, some clear behavioral differences were mentioned. Inbred strains of mice and rats were not developed to reflect these behavioral differences; behavior was studied long after the strains were established.

A somewhat analogous situation seems to exist with regard to domestic animal breeds. With the exception of individual animals or breeders, the various breeds are maintained primarily along lines of morphological or physiological uniformity, with relatively little attention given to behavior. Although there is variation in the behavior of individuals within a breed, there are undoubtedly considerable breed differences.

In livestock species some breed differences in behavior are also recognized. Bulls of dairy breeds are generally less docile than bulls of beef breeds. Systematic observations of mothering behavior in sheep support the assessment of breed-characteristic behavior made by shepherds (Whateley et al., 1974). For example, of six breeds studied, Border-Rommeys were the best mothers in terms of standing with their lambs. Merino ewes were the worst for deserting their lambs, with only 77 percent staying with their lambs at the time of tagging. Shepherds also felt that Border-Rommeys were the best and Merino ewes the worst mothers. Systematic studies on mothering have been carried out on only a few breeds of sheep and almost none of other livestock. It is felt that sows of some pig breeds are more careful not to lie on piglets than sows of other breeds, but definitive observations are missing.

Obviously a great deal could be done on breed-characteristic behavior that would be of great importance in animal production. At the present time, too much faith seems to be placed on the opinions of animal handlers and breeders who have vested interests in seeing their breeds promoted.

An exemplary approach to the analysis of breed differences in behavior that takes the issue out of the hands of those with personal biases is a project by Scott and Fuller (1965) aimed at understanding behavioral differences between five breeds of dogs. The breeds studied were basenjis, beagles, cocker spaniels, Shetland sheepdogs (shelties), and wirehaired fox terriers. A number of tests were devised to measure differences in emotional reactivity, trainability, and problem-solving ability. Emotional reactivity was indicated by distress vocalization, tail wagging, and physiological measures such as heart rate and respiratory rate. The different test situations included an experimenter entering the room with a dog, speaking softly, and leaving. On different occasions he entered the room grabbing the dog's muzzle and forcing the animal's head from side to side. Other tests included sounding a bell or delivering a shock to the animal's legs. By these criteria, the terriers, beagles, and basenjis were significantly more reactive than shelties and cockers. In tests of trainability, the investigators examined the breeds for differences in the degree to which the dog's natural tendencies could be modified or inhibited; that is, how easily the animals became obedient. Examples of the tasks used were leash training and forcing a dog to remain on a stand and then to jump down on command. Cockers were the easiest to train and basenjis and beagles consistently difficult; shelties and terriers varied in that at one task they would be easy to train and in another difficult. A few incidental comments by the investigators are of interest: beagles tended to vocalize excessively, and shelties were difficult to train because they would leap on the handlers.

To examine the dogs under the category of problem-solving behavior Scott and Fuller used several tests. A detour test required the animal to travel around a screen barrier between it and a dish of food. Since the dog could see the food, it was required initially to leave the food in order to get around the barrier. A manipulation test required the dog to manipulate a string to pull a food dish from under a box. Other tests involved trailing a fish smell, discrimination learning, and delayed responding. The dogs of the various breeds were tested between 10 and 22 weeks of age to minimize the influence of prior experience. In most tests there were clear-cut breed differences. The most difficult tasks, such as delayed response, showed the least differences between breeds, but very great differences between individuals within breeds. No single breed performed best in all tasks in which there were clear breed differences. Thus, there was no evidence of a general factor of problem-solving ability or intelligence that differed from one breed to another. In a very general sense, the hunting breeds—the beagles, basenjis, terriers, and cockers—did better than the shelties. The investigators felt that the reason the shelties did less well is that all of the problem-solving tasks were food-motivated and shelties are a breed selected for working flocks of sheep without a food reward.

This investigation by Scott and Fuller is a model for the future analysis of breed differences in behavior in any of the domestic species. It illustrates that general traits such as emotional reactivity and problem-solving ability must be tied down to specific behavioral tests. Obviously, more than one specific test must be used before a judgment can be made with regard to a general behavioral category. Scott and Fuller's study is consistent with the notion that all breeds are about equal in problem solving provided they are adequately motivated and provided that differences in emotional reactivity, sensory capacity, and motor abilities are taken into account. The work of these investigators also illustrates that individual differences, possibly reflecting individual differences in genetic makeup, exist within breeds that are uniform in outward appearance.

Another study at the same laboratory revealed an interesting example of environmental and genetic interactions in the rearing of dogs (Freeman, 1958). The study was concerned with the question of whether indulgent or disciplinary handling of puppies would influence the obedience of the animal at maturity and if there were breed differences in the degree to which the rearing treatment influenced obedience behavior. The breeds studied were shelties, basenjis, wirehaired fox terriers and beagles. Following weaning, half of the pups of each breed were given the indulgence treatment by an experimenter encouraging the pup in play, aggression, or climbing; the pups were never punished. The other half were restrained and taught to sit, stay, and come on command. All pups were then placed in a room with a bowl of meat. Each time a pup ate from the bowl it was punished with a swat on the rump and a shout of "no." The experimenter left the room after three minutes and, through a one-way glass, recorded the time it took before the pup started eating. For shelties and basenjis, the manner of rearing had little influence over their responses; shelties tended to refuse food altogether, and basenjis tended to eat soon after the experimenter left. Beagles and fox terriers, however, were influenced by rearing conditions, in that the indulged pups took longer to return to the food. The principle this experiment demonstrates is that the environmental influence of rearing conditions interacted with the genotype differentially depending on breed.

With companion animals matching the most appropriate animal with a person or family is essential to the well-being of the human-animal relationship. Breed differences in behavioral predispositions, especially among dogs, give us a considerable degree of latitude in choosing the best breed for a person or family. A recent project which utilized forced rankings by 96 expert informants of dog breeds on 13 behavioral characteristics created a data base from which a behavioral profile was developed for each breed (Hart and Hart, 1984). A factor analysis was then performed reducing each breed's score on 13 characteristics to a rank on three factors labeled reactivity, aggressiveness, and trainability. A cluster analysis utilizing the three factors resulted in the creation of seven clusters of dog breeds which could be described by relative rankings on the three factors (see Table 6-1). There are, of course, variations within each breed among

TABLE 6-1 Behavioral profiles of dog breeds: Cluster analysis

Cluster 1	Cluster 2	Cluster 3	Cluster 4	Cluster 5	Cluster 6	Cluster 7
High reactivity, low trainability, medium aggression	Very low reactivity, very low aggression, low trainability	Low reactivity, high aggression, low trainability	Very high trainability, high reactivity, medium aggression	Low aggression, high trainability, low reactivity	Very high aggression, very high trainability, very low reactivity	Very high aggression, high reactivity, medium trainability
Lhasa apso	English bulldog	Samoyed	Poodle (toy)	Newfoundland	German shepherd	Cairn terrier
Pomeranian	Bloodhound	Alaskan malamute	Shih Tzu	Vizsla	Akita	Scottish terrier
Maltese	Norwegian elkhound	Siberian husky	Poodle (standard)	Brittany spaniel	Doberman pinscher	Chihuahua
Cocker spaniel	Old English sheepdog	Saint Bernard	Shetland sheepdog	Labrador retriever	Rottweiler	Schnauzer (miniature)
Boston terrier	Basset hound	Afghan hound	English springer spaniel	Chesapeake Bay retriever		West Highland white terrier
Pekingese		Boxer	Poodle (miniature)	German shorthaired pointer		Airedale terrier
Beagle		Dalmatian	Welsh corgi	Australian shepherd		Fox terrier
Yorkshire terrier		Great Dane	Bichon frise	Keeshond		Silky terrier
Weimaraner		Chow chow		Collie		Dachshund
Irish setter				Golden retriever		
Pug						

individual dogs, but the guidelines expressed in Table 6-1 allow one to take advantage of genetically-related breed differences in behavior in selecting a pet dog. A similar type of analysis of behavioral differences in breeds of cats and horses would be useful as well.

■

APPLIED AND PRACTICAL CONSIDERATIONS

A breeding program designed to alter the behavior of a population or group of animals usually involves the selection of certain individuals as the initial breeding stock. Certain members of their progeny are then selected for breeding the next generation and so on until the desired, or an acceptable, end point of a behavioral phenotype is produced. There are a number of important principles or concepts that should be understood before pursuing such a breeding program.

Selection of Breeding Animals

Animals are usually selected for breeding on the basis of their own behavioral phenotype. However, selection of breeding stock can be based at least as heavily on the behavior of relatives and offspring as on the behavior of the breeding animals themselves. Selection of breeding stock based on examination of progeny or relatives is also a useful procedure when a behavioral pattern under consideration is sexually dimorphic, such as maternal behavior.

The practice of inbreeding is commonly used in breeding programs. Sometimes this is unavoidable because of the smaller number of parents that are selected as breeding stock. There are two main difficulties with inbreeding: it reduces genetic variability along with heritability, thus impairing potential response to selection, and it is likely to greatly increase mortality in young animals and reduce fertility of adults. Specifically, inbred animals are less able to cope with their environment, and more susceptible to diseases and environmental stresses than outbred animals (Losley, 1978; Warwick and Legates, 1979). The mortality stemming from inbreeding is apparent in wild animals as well as domestic ones, and this matter has given cause for warning about the deleterious effects of inbreeding in the genetic management of wild animal populations that are kept in zoological parks (Ralls et al., 1979). A procedure used in some instances is to initiate a breeding program with some inbreeding but then introduce new animals after a few generations have been obtained.

Selection for Certain Behavioral Patterns

Breeding programs may be conducted to enhance desirable characteristics such as some aspect of maternal care in livestock or behavior in cats and dogs. Similarly, breeding programs may be devised to reduce undesirable characteristics such as aggressiveness or excitability. There are two important considerations in this regard. One is that by attending mainly to a particular trait, one may not realize that other aspects of behavior are also being altered. In the example of breeding for better maternal behavior in pigs, one might find that along with showing increased attention to the piglets before lying down, sows displayed an increase in another aspect of maternal responsiveness, namely more aggressive protection of piglets. This might interfere with farm management. It is possible that a compromise must be made between the alteration of a trait of chief concern and the related enhancement of other undesirable traits.

The second consideration, which is related to the first, is that selection for extremes of some behavioral characteristics is likely to produce a decrease in general fitness (Fuller, 1962). This is evident when one considers that both wild and well-established domestic species are the outcome of many generations of selection under certain environmental conditions. There is a balance of physiological and behavioral phenotypes with a high degree of survival success, and this balance may be readily disturbed by rapid and intense selection. It is assumed that selection will not produce new traits but only alter the proportion of desirable and undesirable aspects of existing traits.

The opportunity to enhance a relatively unrecognized trait should not be overlooked. Alcohol preference is not an example of a useful or desirable phenotype in animals, but it does serve as a reminder that some possibly quite useful genetically controlled characteristics may exist to a relatively unrecognized degree, and these traits could be subject to selective breeding.

One of the concepts regarding animal breeding is that certain individuals with highly desirable behavioral phenotypes may reflect a unique genetic combination that cannot be readily fixed by selection. One illustration of this is the principle of noncongruence between genetics and behavior (Fuller, 1962). Assume one is measuring and selecting for some type of learning ability. Next assume that some genetic mechanisms influence learning by affecting emotional reactivity or excitability. If excitability were influenced by four genes in an additive way, then congruence would exist between genotype and excitability. However, if a moderate amount of excitability enhances learning ability but too much impairs it, then the genotype related to excitability would not be additive, it would be noncongruent with respect to learning ability because only a moderate amount of excitability would be most beneficial to learning.

Consideration of Heritability

Since selection is based on behavioral phenotype, efforts to change behavior through selective breeding will be most successful with those traits having a high heritability, in which phenotype is a reliable guide to the underlying genotype. From the elementary representation of heritability, $V_g/(V_g + V_e)$, it is obvious that two factors influence the magnitude of heritability. When the variance attributable to environmental factors (V_e) is small, heritability is greatest. Thus, rearing a breeding stock under the most uniform conditions should yield the largest measures of heritability. At the same time, the greater the genetic variability in a population, the greater the heritability. Generally, V_g is smaller in the more inbred breeds than in heterogeneous groups. Minimizing inbreeding helps maintain a larger heritability throughout the breeding program. Naturally, variability of a behavioral pattern decreases with successive generations of selection during a breeding program and gains in heritability decrease as well. However, at the point where increased heritability is markedly reduced, the breeding program may have produced its intended results.

A particularly low heritability indicates that close attention to various aspects of the environment might yield more appreciable results in the control of behavior. Many breeding programs have been initiated without a proper estimate of heritability. Given some sophistication in behavioral observation and measurement, calculation of realized heritability is a fairly simple matter. The availability of a number of heritability measures of different behavioral patterns of various domestic species would contribute to the field of behavioral genetics and to future breeding programs with domestic animals.

Some traits, such as mothering, are expressed in only one sex, even though we assume both sexes contribute to the trait. In these cases a breeding program is far less successful than when the trait can be measured in both sexes. The same situation arises when the male is unavailable for behavioral observation; for example, when the female dog or cat is allowed to mate with any male in the neighborhood.

References

BOICE, R. 1972. Some behavioural tests of domestication in Norway rats. *Behaviour* 42:198–231.

BOICE, R. 1973. Domestication. *Psychol. Bull.* 80:215–230.

BOREMAN, J., AND E. PRICE. 1972. Social dominance in wild and domestic Norway rats (*Rattus norvegicus*). *Anim. Behav.* 20:534–542.

BROADHURST, P. L., AND S. LEVINE. 1963. Litter size, emotionality and avoidance learning. *Psychol. Rep.* 12:41–42.

BRUELL, J. H. 1967. Behavioral heterosis. In *Behavior-genetic analysis*, pp. 270–286, ed. J. Hirsch. New York: McGraw-Hill.

COLLINS, R. L. 1964. Inheritance of avoidance conditioning in mice: A diallele study. *Science* 143:1188–1190.

CRAIG, J. V., L. L. ORTMAN, AND A. M. GUHL. 1965. Genetic selection for social dominance ability in chickens. *Anim. Behav.* 13:114–131.

FALCONER, D. S. 1960. *Introduction to quantitative genetics.* London: Oliver & Boyd.

FRANK, H., AND M. G. FRANK. 1982. On the effects of domestication on canine social development and behavior. *Appl. Anim. Ethol.* 8:507–525.

FULLER, J. L. 1962. The genetics of behaviour. In *The behaviour of domestic animals*, 2nd ed., pp. 45–64, ed. E. S. E. Hafez. Baltimore: Williams & Wilkins.

GINSBURG, B. E. 1967. Genetic parameters in behavioral research. In *Behavior-genetic analysis*, pp. 135–153, ed. J. Hirsch. New York: McGraw-Hill.

GINSBURG, B. E., AND W. C. ALLEE. 1942. Some effects of conditioning on social dominance and subordination in inbred strains of mice. *Physiol. Zool.* 15:485–506.

GOY, R. W., AND J. S. JAKWAY. 1962. Role of inheritance in determination of sexual behavior patterns. In *Roots of behavior*, pp. 96–122, ed. E. L. Bliss. New York: Harper.

GUHL, A. M., J. V. CRAIG, AND C. D. MUELLER. 1960. Selective breeding for aggressiveness in chickens. *Poultry Sci.* 39:970–980.

HART, B. L., AND L. A. HART. 1984. Selecting the best companion animal. In *The pet connection: Its influence on our health and quality of life*, pp. 180–193, eds. R. K. Anderson, B. L. Hart, and L. A. Hart. Minneapolis: Center to Study Human-Animal Relationships and Environments, University of Minnesota.

LOCKARD, R. 1968. The albino rat: A defensible choice of bad habit? *Amer. Psychol.* 23:734–742.

LOSLEY, J. F. 1978. *Genetics of livestock improvement*. Englewood Cliffs, N.J.: Prentice-Hall.

MANOSEVITZ, M. 1972. Behavioral heterosis: Food competition in mice. *J. Comp. Physiol. Psychol.* 79:46–50.

MASON, J. J., AND E. O. PRICE. 1973. Escape conditioning in wild and domestic Norway rats. *J. Comp. Physiol. Psychol.* 84:403–407.

MC CLEARN, G. E. 1963. The inheritance of behavior. In *Psychology in the making*, pp. 144–252, ed. L. Postman. New York: Knopf.

MC CLEARN, G. E. 1967. Genes, generality, and behavior research. In *Behavior-genetic analysis*, pp. 307–321, ed. J. Hirsch. New York: McGraw-Hill.

MC CLEARN, G, E., AND J. C. DEFRIES. 1973. *An introduction to behavioral genetics*. New York: W. H. Freeman and Company.

MC CLEARN, G. E., AND D. A. ROGERS. 1959. Differences in alcohol preference among inbred strains of mice. *Quart. J. Stud. Alcohol.* 20:691–695.

MC CLEARN, G. E., AND D. A. ROGERS. 1961. Genetic factors in alcohol preference of laboratory mice. *J. Comp. Physiol. Psychol.* 54:116–119.

MC GILL, T. E., AND G. R. TUCKER. 1964. Genotype and sex drive in intact and in castrated male mice. *Science* 145:514–515.

NACHMAN, M. 1959. The inheritance of saccharin preference. *J. Comp. Physiol. Psychol.* 52:451–457.

PRICE, E. O. 1972. Domestication and early experience effects on escape conditioning in Norway rat. *J. Comp. Physiol. Psychol.* 79:51–55.

PRICE, E. O. 1984. Behavioral aspects of animal domestication. *Quart. Rev. Biol.* 59:1–32.

RALLS, K., K. BRUGGER, AND J. BALLOU. 1979. Inbreeding and juvenile mortality in small populations of ungulates. *Science* 206:1101–1102.

ROBERTS, R. C. 1967. Some concepts and methods in quantitative genetics. In *Behavior-genetic analysis*, pp. 214–257, ed. J. Hirsch. New York: McGraw-Hill.

SCOTT, J. P. 1942. Genetic differences in the social behavior of inbred strains of mice. *J. Heredity.* 33:11–15.

SCOTT, J. P., AND J. L. FULLER. 1965.*Genetics and the social behavior of the dog.* Chicago: University of Chicago.

THOMPSON, W. R. 1953.The inheritance of behavior: Behavioral differences in fifteen mouse strains. *Canad. J. Psychol.* 7:145–155.

THOMPSON, W. R. 1967. Some problems in the genetic study of personality and intelligence. In *Behavior-genetic analysis*, pp. 344–365, ed. J. Hirsch. New York: McGraw-Hill.

WAHLSTEN, D. 1972. Genetic experiments with animal learning: A critical review. *Behav. Biol.* 7:143–182.

WARWICK, E. J., AND J. E. LEGATES. 1979. *Breeding and improvement of farm animals.* New York: McGraw-Hill.

WHATELEY, J., R. KILGOUR, AND D. C. DALTON. 1974. Behaviour of hill country sheep breeds during farming routines. *Proc. N.Z. Soc. Anim. Prod.* 34:28–36.

WEIR, M. W., AND J. C. DEFRIES. 1964. Prenatal maternal influence on behavior in mice: Evidence of a genetic basis. *J. Comp. Physiol. Psychol.* 58:412–417.

WILLHAM, R. L., D. F. COX, AND G. G. KARAS. 1963. Genetic variation in a measure of avoidance learning in swine. *J. Comp. Physiol. Psychol.* 56:294–297.

CHAPTER SEVEN ▪ EARLY EXPERIENCE AND BEHAVIOR

Animal keepers have a profound influence on the behavior of young animals because we have easy access to them and extensive control over them. Anyone familiar with child psychology realizes the tremendous impact that early experiences can have on a person's later behavior. Environmental and experiential factors may have such profound effects on young organisms that many physiological as well as behavioral capabilities of the adult are changed through early experience. For the most part, the capabilities are relatively difficult to alter later in an animal's life. When an animal's sensory systems first start to function, the first stimuli it receives obviously have a greater impact than experiences of a comparable type later in life. One of the reasons some early experiences are almost irreversible is that the brain is rapidly developing at the time of birth and in the neonatal period (Figure 7-1). Outside stimuli can have an effect on brain development and the effects become virtually "locked" into the brain.

A consideration of the effects of early experience focuses on species in which the young are born immature. In these species there is a prolonged period over which the sensory and

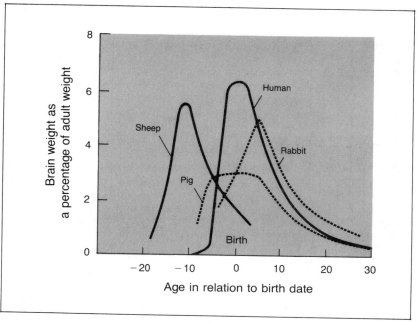

Figure 7-1 This graph illustrates the brain growth spurts of sheep, pigs, rabbits, and humans. The gain in brain weight is expressed as a percentage of adult weight for the following units of time: human, months; pig, weeks; sheep, five days; rabbits, two days. (From Dobbing and Sands, 1979.)

experiential factors may act on the organism to influence the developing brain. For example, isolation in early life has more profound effects on altricial than precocial species.

The dramatic effects of early experience that are often cited to illustrate certain principles and concepts involve gross deprivation or severe stress. An example of the type of extreme effects that early experiences may have is given by research showing that when some sensory systems are rendered completely nonfunctional from the time of birth, there are physiological and anatomical deficiencies in the parts of the brain dealing with these systems. Visual deprivation results in abnormal development in the retina and a reduced size of structures along the visual pathway in the brain (Hubel, 1979). Effects of early sensory experiences on brain anatomy have been studied extensively in rats. Animals raised in bland, impoverished environments have been compared with rats raised in an enriched environment with plenty of toys, runways, and tunnels to explore and other rats with which to interact. Rats raised in the enriched environment have a thicker cerebral cortex, increased synaptic contacts between neurons, wider dendritic branching of neurons, and higher levels of some important neurotransmitters (Fiala et al., 1978; Rosenzweig and Bennett, 1977).

■

PRENATAL AND NEONATAL INFLUENCES

When one considers that the brain of the unborn mammal in late gestation and in early postnatal life is quite developed but still undergoing some maturation and growth, it is easy to understand how stimuli which we normally consider to be rather mild can have profound effects on an animal's physiological capacities and behavior even as an adult. At no other time in the animal's life is the brain again in this formative state where stimuli that impinge upon it can affect the permanent structure and functioning. Admittedly, in these early stages of life, the organism is not exposed to a great many stimuli. However, blood-borne chemicals can get to the brain of the developing fetus through the maternal vascular system and the newborn animal can be affected by changes in temperature, handling, and by some olfactory and auditory stimuli. We will examine some of these effects below.

Effects of Blood-Borne Chemicals on the Fetus

In the uterine environment the organism is rather well insulated from outside stimuli. However, the nervous system of a fetus can detect kinesthetic stimuli from maternal movement and chemical stimuli such as hormones produced by its own endocrine glands or hormones that cross the placenta from maternal circulation.

Experiments on rats and mice have demonstrated that in some instances, subjecting a pregnant mother to repeated aversive stimuli such as electrical shocks and making her continually fearful may result in offspring that are excessively emotional as adults. The emotionality that characterizes the behavior of the young rats includes freezing and frequent elimination in an open field. This effect is presumably the result of increased secretion of adrenal hormones in the pregnant female and the transfer of these hormones, by the placenta, to the fetus. Additional experiments have suggested that making a female continuously fearful even prior to conception may increase the emotionality of offspring that are conceived later (Joffe, 1965).

The effect of prenatal stress on animals other than rodents has yet to be determined. Even with rodents, such as the house mouse, there is now evidence that this effect of prenatal stress occurs in only certain genetic strains (Weir and DeFries, 1964).

One other aspect of prenatal experience that has received some attention is the role of maternal stress on the reproductive function of the offspring. Male rats of mothers exposed to stress during the latter one-third of gestation display low levels of male copulatory behavior and sometimes high rates of feminine sexual behavior (Dahlof et al., 1977; Ward, 1972). It appears as though stress of the mother alters the normal pattern of gonadal testosterone secretion in neonatal males so that they do not undergo the same fetal masculinizing process as normal males (Ward and Weisz, 1980). Additional research has shown that female rats from prenatally stressed mothers tend to have a high rate of reproductive failure. This effect is also attributed to the activation of the mother's adrenal cortex and secretion of androgens that affect developing females (Herrenkohl, 1979).

Effect of Handling on Neonatal Animals

We have a tendency to believe that an organism develops optimally in an environment that is constant in terms of temperature, comfort, and freedom from aversive stimulation. Through a good deal of research on rats, we now have a basis for concluding that some degree of stress in the neonatal period of altricial animals accelerates body growth and reduces emotionality. One of the noteworthy effects of neonatal stress is evidence of increased resistance to some diseases.

In a typical experiment, groups of infant rats might be given one of three different types of daily treatment: mild electric shock in a box, just being handled and placed in a box, or remaining undisturbed. In later tests for emotionality in an open field, both the handled and the shocked rats invariably move about more freely, defecate and urinate less frequently, and show little sign of emotional activation. The nonmanipulated animals tend to crouch or freeze in a corner and defecate and urinate frequently (Denenberg, 1962; Levine, 1969). Since the initial observations on this phenomenon, it has been learned that other types of treatment of young rats, such as mechanical shaking or brief exposures to a cold environment, can have the same effect as handling. The effect of infant handling in reducing emotionality and increasing activity in novel environments has been found in virtually all altricial species tested, including rabbits, cats, and various species of rodents (Denenberg et al., 1973; Wilson et al., 1965).

Rats that are manipulated during the neonatal period generally have a more rapid rate of development of many organ systems. There is accelerated maturation of the central nervous system, the eyes open sooner, motor coordination develops earlier, hair growth occurs sooner, and the animals gain weight faster. Also, the adrenal glands are smaller, which is probably related to a decreased tendency toward emotionality. Some authorities consider these effects as making the young healthier and point to other signs that handled animals are healthier. Handled rats, for example, reportedly survive stressful water and food deprivation better than nonhandled controls (Levine and Otis, 1958; Denenberg and Karas, 1959).

Increased Resistance to Certain Diseases   Another area of the handling phenomenon that has concerned investigators is resistance to certain diseases. Although the suggestion that neonatal stress may enhance an animal's resistance to certain diseases is potentially very important, it is difficult to study this area because of the problem of producing identical disease conditions in experimental and control groups. One condition that is fairly replicable is gastric ulceration in rats produced by a conflict situation. Handling of rats in the first few days of life increases the resistance to such ulcers (Ader, 1965; Schaefer and Darbes, 1972).

More extensive studies of resistance to disease have been conducted in mice. Analysis of susceptibility to transplanted tumors, leukemias, and encephalomyocarditis, for example, has shown that resistance in some instances is increased by neonatal handling whereas resistance to other diseases is decreased or not influenced at all. It appears that the effect of neonatal handling in altering susceptibility to disease is dependent on the particular disease (Daly, 1973; Friedman et al., 1969). The question of handling effects on disease resistance could be extremely important, and it is unfortunate that most of the studies have involved the mouse as the experimental model. Physical stimulation probably has a more severe effect on the mouse than on the rat or carnivores because the mouse is a more rapidly developing organism.

Physiological Mechanism of the Handling Effect Related to the question of animal health and growth is the development and functioning of the pituitary-adrenal cortex axis, because this system plays an important role in resistance to disease and response to stress. It is established that neonatally handled rats have as adults a low adrenal corticosteroid output in novel but harmless situations. Furthermore, their adrenal corticoid output seems to match the actual demands. Under actual physical stress, such as they might encounter in a predator attack, a brief high output of corticosteroid hormones in these animals temporarily accelerates metabolism, mobilizes energy stores, and prepares them for physical exertion. When the stress is over, the adrenal cortex quiets down. Animals that have not been handled tend to have a corticosteroid reponse that is prolonged in a harmless novel environment as well as under actual stress. This prolonged secretion pattern works to the animal's disadvantage by exhausting its resources, impairing immune reactions, and increasing its susceptibility to infection.

The mechanism by which the corticosteroid output is influenced by neonatal handling is believed to be related to the concept that arousal of corticosteroid secretion in neonatal animals acts on the infant brain to promote a type of neural development that allows a release of corticosteroids matching the actual stress of the environment (Denenberg and Zarrow, 1971; Levine and Mullins, 1966). A problem with this hypothesis is that only handling, not other types of stress in infancy, produces this effect on corticosteroid secretion (Pfeifer, et al., 1976).

Since handling of neonatal rats and other animals seems to have a variety of behavioral and physiological influences, it seems logical to expect that possibly several physiological factors may be operating on the infant to produce these effects, and that influences on the pituitary-adrenal-corticosteroid axis may be only one part of the total picture. Some studies indicate that the effects of handling may be brought about through maternal influences. Handled pups, being colder, evoke different maternal behavior than pups that are always warm. It has been suggested that perhaps a combination of tactile stimulation, hypothermia, stress, and maternal influences all play a role in the production of the general physiological effects of early stimulation (Russell, 1971).

Handling Effects in Wild and Domestic Animals   One of the issues regarding the early handling syndrome is whether a certain amount of stress and hypothermia during infancy is normal in wild rodents. It has been argued that the nests of wild mice and rats are very stable in temperature and humidity, and the neonates normally experience a very consistent environment. If neonatal stress does not occur in nature in rodents, one might argue that the effects of handling are not necessarily adaptive. There is no particular reason, for example, to believe that reduction of emotionality and accelerated body growth are adaptive in the wild (Daly, 1973).

Whether the neonatal offspring in wild canids and felids experience environmental disturbance equivalent to the handling effect is an open question. Their neonatal environment is probably less constant than that of rodents. On the domestic scene, we frequently do disturb neonatal dogs and cats, and the effects on their behavior as pets appear to be desirable. Kittens handled from birth through 45 days of age approach strange objects and adults more readily than controls (Wilson et al., 1965). Siamese kittens that are handled for 10 minutes a day for the first 30 days of life have accelerated development. Their eyes open a day or so sooner, the typical Siamese coloration appears sooner, and they emerge from the nest an average of three days sooner (Meier, 1961).

■

MATERNAL AND PEER INFLUENCES: EFFECTS OF DEPRIVATION

Under natural conditions mothers spend virtually 24 hours a day with their infants. If the infants have littermates, they are also with their littermates 24 hours a day at least until the time of normal weaning. It is only natural to expect that disruption of this around-the-clock maternal and littermate contact can have profound influences. Experimental work has shown that behavioral and physiological functions, as well as disease resistance, can be impaired by various kinds of maternal and litter-mate deprivation.

Behavioral Effects of Deprivation

Unfortunately few controlled studies have been done on the effects of maternal deprivation in domestic animals. The effects that have been noticed are not the severe behavioral pathologies reported for monkeys, but more subtle changes in emotional behavior. Infant monkeys, for example, that are isolated soon after birth and never allowed to interact with their mothers or other young monkeys until they are near adulthood may engage in stereotyped crouching, self-clasping, body rocking and head banging and self-mutilation (Berkson 1967; Harlow and Harlow, 1962).

Kittens separated from their mothers at the normal weaning age of six weeks occasionally cry for a day or two but otherwise seem to suffer little stress. Kittens separated at 12 weeks cry even less and show practically no indication of stress from the separation. Kittens separated from their mothers when they are two weeks of age cry intensely for as long as a week or more following the separation. The kittens separated at two weeks are more suspicious, cautious, and aggressive as adults. They exhibit more hissing and scratching toward experimenters than cats weaned later. In quantitative tests of behavioral effects, it was found that kittens isolated at two weeks were the most randomly active in a simple learning task and were the slowest to learn. When run through a feeding frustration task in which they received electrical shocks after opening the food container, the animals isolated at two weeks developed an asthmalike condition characterized by wheezing and sneezing that could not be related to an obvious respiratory disorder (Seitz, 1959).

Since dogs are so social, one might expect the effects of complete maternal deprivation in dogs to be quite devastating. However, the behavioral effects are rather moderate. Dogs isolated from each other from early in life have been reported to be very subordinate to other dogs, less competitive, more withdrawn, and fearful of novel objects (Thompson and Melzack, 1956). Fox and Stelzner (1966) found roughly the same behavioral effects with isolation of puppies at four to five weeks of age.

In summarizing, it is probably fair to say that early isolation from both the mother and littermates impairs adult social behavior. Behavioral abnormalities such as self-mutilation or compulsive stereotypy are uncommon. In dogs the antisocial behavior and timidity are probably important enough to render the animals less satisfactory as pets than normally reared dogs. In cats, isolation tends to make them less social toward people. Such a cat released into the wild might not be capable of responding appropriately to members of the opposite sex during the breeding season or capable of fighting to defend a territory or nest. Research shows that in general, the earlier the isolation, the more pronounced the impairment in social behavior.

One would expect the effects of isolation from the mother on precocial species such as cattle, sheep, swine, and horses to be less apparent than in cats and dogs, and this is what has been seen. In fact, dairy calves may never see their mothers, yet as adults they are considered behaviorally normal. Beef calves, foals, lambs, and piglets, however, are normally left with their dams until weaning. The young normally experience interactions with their mothers and play with peers. A few experiments have involved isolating young ungulates at the time of birth. Calves that are group reared with other calves after being isolated from their mothers at one day of age are dominant over calves reared in isolation. The isolates were found to go off by themselves more often than group-reared calves (Broom and Leaver, 1978).

Lambs that are raised in isolation from other lambs respond to strange lambs when tested later, but only after prolonged adaptation to the strange lambs. Lambs reared in a group without mothers interact much more freely with strange lambs (Zito et al., 1977).

Although beef calves, lambs, and piglets that are intended for sale soon after weaning are allowed to run with their mothers, farmers may separate males that they intend to use for breeding as early as three weeks of age. As mentioned, the effects of isolation on social behavior are not terribly dramatic. However, minor influences may be quite important in sexual behavior. Boars that were reared under isolated conditions from three weeks of age suffered depressed copulatory performance compared with those reared in an all-male or mixed-sex group. The isolation effects are not restricted to early life. Isolating adult boars from female pigs for 12 weeks also reduces copulatory performance in comparison to boars housed near females (Hemsworth, 1980).

In most of the studies on the isolation of farm animals, there is no indication that isolation reduces growth rate or weight gain. One factor that has not been examined in either altricial or precocial species is whether early separation from the mothers and peers may predispose the young to certain diseases by lowering body resistance or immunity. Work on the laboratory rat suggests this is an area in need of examination.

Physiological Effects of Early Separation

Representative of the several studies on early separation in rodents is one in which it was found that rats weaned at 15 days of age developed significantly more ulcers from conflict situations than controls weaned at the normal time of from 22 to 30 days. Rats weaned at 15 days of age were also more susceptible to a transplanted tumor than those weaned at the usual time (Ader and Friedman, 1965). Rabbits isolated from littermates starting at birth had a slower rate of weight gain, were less efficient in utilizing ingested food, and exhibited a disruption of the normal sleep pattern (De Santis et al., 1977). Early isolation does not always lead to impaired disease resistance. Some studies have found that early weaning actually can lead to enhanced resistance to some diseases (LaBarba et al., 1970).

DEVELOPMENT OF ATTACHMENTS AND PREFERENCES

On the domestic scene we find that some animals appear to acquire strange playmates: sheep sometimes prefer the company of goats; dogs may be more closely attached to people than other dogs; cats enjoy playing with dogs. In a classic experiment newborn kittens raised in the same cage with a rat or a sparrow became so strongly attached to their cage mates that they developed a hostile attitude toward other kittens that came near the cage (Kuo, 1967).

General Attachment Effects

The influence of early experience in promoting attachments is quite evident in both livestock and companion animals. About the only way, for example, that cats will interact freely with dogs, to the point of engaging them in rough-and-tumble play, is if they are raised with a dog from early kittenhood. Dogs and cats that take well to automobile trips and grooming are usually those animals that were exposed to these experiences very early.

In livestock animals attachments between species are usually a function of early prolonged exposure of animals to each other. Cattle, sheep, and goats may become overly attached to human handlers, especially if they are raised in pens isolated from conspecifics and fed and played with by people. The attachment to people may be so strong that even sexual responses are made toward people (Figure 7-2) (Sambraus and Sambraus, 1979). Sheep and goats have strong herding tendencies with regard to their own species. However, if raised and exposed only to members of a different species, a lamb or kid may prefer to associate with the species with which it was raised (Cairns, 1966a; Hersher et al., 1963; Tomlinson and Price, 1980). Orphaned lambs are occasionally raised by bottle by human handlers, and here too, the young lambs may become strongly attached to their foster species. Behaviorists have noted that attachments are more easily made to some objects than others. For example, attachments are almost always made to animate and not inanimate objects. Cairns (1966b) and others have noted the attachment behavior is guided by the type of stimulus and the length of association.

These attachments are not irreversible. Tomlinson and Price (1980) noted that even when sheep or goats were reared without even seeing a member of their own species, they would develop an affinity for conspecifics as adults if left with conspecifics for one to two months.

Imprinting and Critical Periods

A classic example of early attachment is the phenomenon of imprinting in newly hatched precocial birds such as ducks or chickens. Konrad Lorenz was the first to note that baby geese hatched by a mother goose always followed her around the barnyard, whereas baby geese hatched by incubators tended to follow an object in their environment that was one of the first things they saw after hatching (Lorenz, 1937). Young goslings raised in the incubator and imprinted onto Lorenz would follow him even when they had an opportunity to follow a mother goose instead. Lorenz also noted that baby goslings that were im-

Figure 7-2   Farm animals are sometimes quite prone to making sexual responses toward people; this often happens when they are bottle-raised in close contact with people. They may form stronger attachments to people than to members of their own species. (From Sambraus and Sambraus, 1975).

printed onto him tended to make sexual advances toward him after they became adults.

Imprinting was later studied in more detail by Hess (1973). Objects presented to the ducklings before and after this critical period were presumably much less effective in inducing the imprinting reaction. It was also determined that allowing ducklings to see and follow an environmental object for only a few minutes within the three-hour critical period was sufficient for imprinting to take place.

In the restricted sense the term imprinting is usually reserved for strong attachments that presumably occur within a very short time span, are relatively irreversible, and are formed in a particular development period of an animal's life. Most studies of imprinting have addressed how early experiences affect the young's choice of a mother figure. The relationship of this maternal imprinting to sexual imprinting, which may occur at a different age in life, has not been clarified (Brown, 1975).

The terms critical period, sensitive phase, vulnerable point, and so forth imply that a given event occurring at one stage in an animal's life has stronger influences than it would at other times. We see plenty of examples in physiological processes of critical periods. Masculinizing effects of androgens on animals are found only at certain stages, protein malnutrition has permanent stunting effects only early in life, and the influence of visual experience on development of the visual system is seen only at a certain developmental stage. It is of interest to think of critical periods with regard to behavioral processes in animals. We do not like the idea, however, of a human child missing the developmental bus by not being treated in a certain way at a specific time (Clarke and Clarke, 1976).

Do absolute critical periods exist for behavioral processes such that if a young animal is dealt an unfortunate hand in life experiences, rehabilitation as an adult is hopeless? The answer appears to be an unqualified no. The more behaviorists study so-called critical period phenomena, the more vague and nonexistent the boundaries of the critical periods become. Therefore, the term sensitive period is used more frequently for such phenomena nowadays.

The debate in this area has tended to center around imprinting in birds as a model for all critical periods. One of the critics in this area, P. Bateson (1979), summarizes the available evidence about critical periods by making an analogy to a train traveling one way from a place called conception to a place where it disappears off the tracks. The train starts with all the windows closed. The critical period hypothesis would have the train's windows thrown open and the passengers exposed to the outside world, and then a little later shut again. For the purpose of analogy to animals, the windows in different compartments could be opened and closed at different times for different critical periods.

As an alternative to this brief opening of windows it is possible to imagine that the windows, once open, remain open. Passengers in each compartment may or may not attend to the outside world depending on competing activities or what they have seen before. With regard to imprinting Bateson concludes that the animal is simply occupied sufficiently with the object to which it has imprinted that it ignores other stimuli (the windows of the train remain open but the passenger is not looking out).

One of the more vivid documentations of this veiwpoint is Mason and Kenney's (1974) work in getting young rhesus monkeys to form strong attachments to dogs as surrogate mothers (Figure 7-3). Monkeys raised with cage mates or with their own mothers for three months went on to form strong specific bonds with a companion dog when taken from their previous conspecific companions and gradually introduced to the dog. The dogs and young monkeys played together and groomed each other, the dogs by licking and the monkeys by using their fingers. If the dogs were taken away, the young monkeys vocalized, paced, and attempted to escape from the cage, presumably to get to their canine companions.

Socialization in the Dog

A particularly elaborate conceptualization of the early attachment effect has been developed by Scott and Fuller (1965) in their work on the socialization of the dog. In theory, the development of social attachments in the dog is probably not much different than what we have seen with sheep or cattle. But because people customarily interact more intimately with their dogs than with large animals there has been particular interest in the effects of a dog's early experience on its behavior as a pet toward people. Also, the dog, being more altricial in development than sheep and cattle, lends itself to a longer analysis of the optimal times for influencing attachments and social responses.

In the neonatal period, extending from birth to two weeks of age, a puppy's olfactory, thermal, and tactile senses are functioning, but the puppy is isolated from visual and auditory inputs of the external world. The pup's behavior is mostly restricted to

Figure 7-3 The attachment of this young monkey to this dog occurred after the monkey had already formed a previous attachment to its mother. Thus in monkeys, as in most other species, the early attachment bonds are not irreversible if suitable substitutions are found for later attachment. (Photograph from Mason and Kenney, 1974.)

sleeping and nursing. The neonatal period phases into the transition period at two to three weeks of age when the eyes and ears begin to function, and the puppy begins to move about more readily.

Starting at about three weeks of age and extending through weaning until about 12 weeks of age, puppies are maximally susceptible to socializing influences. Although dogs have genetically acquired predispositions for social responses, these responses must be shaped and refined by social interactions. If exposed to people during the socialization phase, dogs shape their social responses to include people basically as a conspecific. Generally they accept a subordinate role to most people. If taken from the litter just after the transition period and only exposed to human handlers through 12 weeks of age, puppies are likely to show inadequate social responses to other dogs. They may fight too readily or be extremely withdrawn and nervous around other dogs. Male dogs isolated from other dogs during this time may never be normal in their sexual interactions with female dogs. In one investigation it was found that such males played excessively with females and did not display normal sexual mounts (Beach, 1968).

If during the socialization period puppies have very little contact with people, their behavior as adults will be predominantly characterized by fear and escape responses toward people. They may never act relaxed around people and may not use canine social responses, such as submissive behavior, toward them.

■

APPLIED AND PRACTICAL CONSIDERATIONS

Many of the concepts examined in this chapter have involved work on the laboratory rat or on just one of the domestic species. It is sometimes difficult to make meaningful extrapolations from such limited data, but the implications for raising animals are so great that such extrapolations seem warranted. The early life of domestic animals is subjected to about as much human intervention as that of children. Sometimes this intervention is in the way of mistreatment, but more frequently, it is in the way of isolation.

Prenatal and Neonatal Stress

Because of the limited amount of experimental work in the area of prenatal influences, this is a difficult area for which to draw sound conclusions. Clearly, stress of a pregnant female has different effects on the young than stress of the young after birth. In rodents, stress of a pregnant female has often resulted in excessive emotionality and impairment of reproductive function. Given these observations it can probably be concluded that stressful stimuli resulting from transportation, punishment, or

even a change in home environment should be avoided in females that are pregnant. In dogs and cats the latter part of pregnancy is going to be the most critical. In ungulates which are
born in a much more developed stage than carnivores, the crucial time would be more in the middle of pregnancy.

The studies on neonatal handling have implications for the
raising of dogs and cats. It appears as though moderate stress,
such as handling, during the first few days of life is not detrimental and could be beneficial from both health and behavioral
standpoints. One might expect a reduction of emotionality,
more rapid development of organ systems, and possibly greater
resistance to some diseases in handled animals compared with
dogs and cats raised in the most consistent environment. A reduction in emotional reactivity would be desirable, but it could
be overdone, and work on mice and rats shows that excessive
neonatal handling may result in insufficient emotionality for optimal learning of some tasks.

Most neonatal dogs and cats probably receive enough incidental disturbance that handling need not be intentionally
done in order to produce the beneficial effects. On the other
hand, there is no reason for dog and cat breeders to take a
"hands-off" approach for fear that handling will have detrimental effects. Procedures such as surgical removal of dew claws or
docking of tails, which are routine for some breeds, should be
performed in the neonatal period. The stress of these operations
will be the least disruptive at this time.

The handling concept probably does not hold for precocial
animals such as cattle, sheep, and swine. Even if the handling
concept were applicable, the young ungulate probably needs no
intentional extra handling since the usual farm situation, with
daily changing temperatures and movement from one stall or
pen to another, undoubtedly provides adequate stress.

Maternal Deprivation

One of the most frequent practices in raising domestic animals is early weaning and separation of the young from the dam.
Raising orphaned dogs and cats with no maternal contact is not
uncommon. As with the studies on handling, we are forced to
extrapolate somewhat from studies on rodents, in which early
weaning produces changes in emotionality and may decrease an
animal's resistance to disease or stress. Limited work on cats

reveals that cats weaned early are more cautious, apprehensive, and aggressive as adults. Dogs tend to be more withdrawn and timid. As adults the animals may not show normal sexual behavior.

Unless one is very conscientious about handling and stimulating orphan dogs and cats, the early life of such animals represents a form of sensory and affectional deprivation. It is, of course, impossible to replace a mother, and hence all animals raised alone as bottle-fed orphans are deprived of some early experience. Early contact and interaction with other animals should be provided to orphans by leaving all the littermates together. With just one orphaned infant, it seems advisable to attempt to introduce it into another litter rather than raise it alone.

Development of Attachments

Lambs, goats, and calves form very strong attachments to people if they are raised on the bottle by people. These attachments can stand in the way of using such animals in a breeding herd since they may be more human-oriented than animal-oriented.

The attachment and socialization concept is already widely applied in raising dogs. It is believed that the best time to bring puppies into a new home is at six to eight weeks of age. Obtaining a pup before this time may result in inadequate socialization with other dogs as a result of insufficient opportunity for it to interact with littermates. Dogs may become overly attached to human beings to the extent that they consistently avoid other dogs and even make sexual advances toward people. Adopting puppies after about 12 weeks may lead to inadequate attachment to and socialization with people if the dogs have little exposure to people before this time. Such dogs may be difficult to control or dominate.

Perhaps the most practical approach related to early experience in juvenile animals is to attempt to shape an animal's behavior for a particular role on a farm or in a home. If a horse is to be transported, exposure to stimuli involved in transportation should begin at an early age. If dogs are to live in a family with young children, they should be exposed to young children early in life. If one wants a cat to get along well with dogs, the cat should be constantly exposed to a dog at an early age.

References

ADER, R. 1965. Effects of early experience and differential housing on behavior and susceptibility to gastric erosions in the rat. *J. Comp. Physiol. Psychol.* 60:233–238.

ADER, R., AND S. B. FRIEDMAN. 1965. Differential early experience and susceptibility to transplanted tumor in the rat. *J. Comp. Physiol. Psychol.* 59:361–364.

BATESON, P. 1979. How do sensitive periods arise and what are they for? *Anim. Behav.* 27:470–486.

BEACH, F. A. 1968. Coital behavior in dogs. III. Effects of early isolation on mating in males. *Behaviour* 30:218–238.

BERKSON, G. 1967. Abnormal stereotyped motor acts. In *Comparative psychopathology*, pp. 76–94, eds. J. Zubin and H. F. Hunt. New York: Grune & Stratton.

BROOM, D. M., AND J. D. LEAVER. 1978. Effects of group-rearing or partial isolation on later social behaviour of calves. *Anim. Behav.* 26:1255–1263.

BROWN, J. L. 1975. *The evolution of behavior.* New York: Norton.

CAIRNS, R. B. 1966a. Attachment behavior of mammals. *Psychol. Rev.* 73:409–426.

CAIRNS, R. B. 1966b. Development, maintenance and extinction of social attachment behavior in sheep. *J. Comp. Physiol. Psychol.* 62:298–306.

CAIRNS, R. B. 1972. Attachment and dependency: A psychobiological and social learning synthesis. In *Attachment and dependency*, pp. 29–80, ed. J. L. Gewirtz. Washington, D.C.: Winston.

CLARKE, A. M., AND A. B. CLARKE. 1976. *Early experience: Myths and evidence.* London: Open Books.

DAHLOF, L., E. HÄRD, AND K. LARSSON. 1977. Influence of maternal stress on offspring sexual behaviour. *Anim. Behav.* 25:958–963.

DALY, M. 1973. Early stimulation of rodents: A critical review of present interpretations. *Brit. J. Psychol.* 64:435–460.

DENENBERG, V. H. 1962. The effects of early experience. In *The behaviour of domestic animals*, pp. 95–130, ed. E. S. E. Hafez. Baltimore: Williams & Wilkins.

DENENBERG, V. H., AND G. G. KARAS. 1959. Effects of differential infantile handling upon weight gain and mortality in the rat and mouse. *Science* 130:629–630.

DENENBERG, V. H., M. V. WYLY, J. K. BURNS, AND M. X. ZARROW. 1973. Behavioral effects of handling rabbits in infancy. *Physiol. Behav.* 10:1001–1004.

DENENBERG, V. H., AND ZARROW, M. X. 1971. Effects of handling in infancy upon adult behavior and adrenocortical activity: Suggestions for a neuroendocrine mechanism. In *Early childhood: The development of self-regulatory mechanisms*, pp. 48–64, eds. W. W. Walcher and D. L. Peters. New York: Academic.

DESANTIS, D., S. WAITE, E. B. THOMAN, AND V. H. DENENBERG. 1977. Effects of isolation rearing upon behavior state organization and growth in the rabbit. *Behav. Biol.* 21:273–285.

DOBBING, J., AND J. SANDS. 1971. Comparative aspects of the brain growth spurt. *Early Human Dev.* 31:79–83.

FIALA, B. A., J. N. JOYCE, AND W. T. GREENOUGH. 1978. Environmental complexity modulates growth of granule cell dendrites in developing but not adult hippocampus of rats. *Experimental Neurol.* 59:372–383.

FOX, M. W., AND D. STELZNER. 1966. Behaviour effects of differential early experience in the dog. *Anim. Behav.* 14:273–281.

FRIEDMAN, S. B., L. A. GLASGOW, AND R. ADER. 1969. Psychosocial factors modifying host experience to experimental infections. *Ann. N.Y. Acad. Sci.* 164:381–393.

HARLOW, H. F., AND M. K. HARLOW. 1962. Social deprivation in monkeys. *Sci. Amer.* 207:137–146.

HEMSWORTH, P. H. 1980. The social environment and the sexual behaviour of the domestic boar. *Appl. Anim. Ethol.* 6:306.

HERRENKOHL, L. R. 1979. Prenatal stress reduces fertility and fecundity in female offspring. *Science* 206:1097–1099.

HERSHER, L., J. B. RICHMOND, AND A. U. MOORE. 1963. Modifiability of the critical period for the development of maternal behaviour in sheep and goats. *Behaviour* 20:311–320.

HESS, E. H. 1973. *Imprinting.* New York: Van Nostrand.

HUBEL, D. H. 1979. The visual cortex of normal and deprived monkeys. *Amer. Scient.* 67:532–543.

JOFFE, J. M. 1965. Genotype and prenatal and premating stress interact to affect adult behavior in rats. *Science* 150:1844–1845.

KUO, Z.-Y. 1967. *The dynamics of behavior development—an epigenetic view.* New York: Random House.

LA BARBA, R. C., J. L. WHITE, J. LAZAR, AND M. KLEIN. 1970. Early maternal separation and the response to Ehrlich carcinoma in BALC/c mice. *Devel. Psychobiol.* 3:78–80.

LEVINE, S. 1969. Infantile stimulation: A perspective. In *Stimulation in early infancy*, pp. 3–19, ed. A. Ambrose. London: Academic.

LEVINE, S., AND R. F. MULLINS. 1966. Hormonal influences on brain organization in infant rats. *Science* 152:1585–1592.

LEVINE, S., AND L. S. OTIS. 1958. The effects of handling before and after weaning on the resistance of albino rats to later deprivation. *Can. J. Psychol.* 12:103–108.

LORENZ, K. 1937. The companion in the bird's world. *Auk* 54:245–273.

MASON, W., AND M. KENNEY. 1974. Redirection of filial attachments in rhesus monkeys: Dogs as mother surrogates. *Science* 183:1209–1211.

MEIER, G. W. 1961. Infantile handling and development in Siamese kittens. *J. Comp. Physiol. Psychol.* 54:284–286.

PFEIFER, W. D., R. ROFUNDO, M. MYERS, AND V. H. DENENBERG. 1976. Stimulation in infancy: Unique effects of handling. *Physiol. Behav.* 17:781–784.

ROSENZWEIG, M. R., AND E. L. BENNETT. 1977. Experiential influences on brain anatomy and brain chemistry in rodents. In *Studies on the development of behavior and the nervous system*, vol. 4, *Early influences*, pp. 289–327, ed. G. Gottlieb. New York: Academic.

RUSSELL, P. 1971. "Infantile simulation" in rodents: A consideration of possible mechanisms. *Psychol. Bull.* 75:192–202.

SAMBRAUS, H. H., AND D. SAMBRAUS. 1975. Prägung von Nutztieren auf Menschen. *Z. Tierpsychol.* 38:1–17.

SCHAEFER, G. J., AND A. DARBES. 1972. The effects of preweaning handling and postweaning housing on behavior and resistance to deprivation induced stress in the rat. *Devel. Psychobiol.* 5:231–238.

SCOTT, J. P., AND J. L. FULLER. 1965. *Genetics and the social behavior of the dog*. Chicago: University of Chicago.

SEITZ, P. 1959. Infantile experience and adult behavior in animal subjects. *Psychosom. Med.* 21:353–378.

THOMPSON, W. R., AND R. MELZACK. 1956. Early environment. *Sci. Amer.* 194:38–42.

TOMLINSON, K. A., AND E. O. PRICE. 1980. The establishment and reversibility of species affinities in domestic sheep and goats. *Anim. Behav.* 28:325–330.

WARD, I. 1972. Prenatal stress feminizes and demasculinizes the behavior of males. *Science* 175:82–84.

WARD, I. L., AND J. WEISZ. 1980. Maternal stress alters plasma testosterone in fetal males. *Science* 207:328–329.

WEIR, M. W., AND J. C. DE FRIES. 1964. Prenatal maternal influence on behavior in mice: Evidence of a genetic basis. *J. Comp. Physiol. Psychol.* 58:412–417.

WILSON, M., J. M. WARREN, AND L. ABBOT. 1965. Infantile stimulation, activity, and learning by cats. *Child Devel.* 36:843–853.

ZITO, C. A., L. L. WILSON, AND H. B. GRAVES. 1977. Some effects of social deprivation on behavioral development of lambs. *Appl. Anim. Ethol.* 3:367–377.

CHAPTER EIGHT ■ ANIMAL LEARNING

I f you watch an animal during one of its active periods, you will observe an integrated mixture of innate responses and learned behavior. When a horse, for example, moves from one location of a pasture to a new one, the direction of movement may reflect a learned response acquired on the basis of previous food or water reward. The animal may have learned to move to the new location at a specific time of day. While in the process of moving, the horse may also interact with conspecifics using species-typical, innate responses, such as submission or threat. Actually, the choice of responses, whether threat or submission, is learned dependent on the outcome of previous aggressive encounters. Although it is quite arbitrary and artificial to distinguish between learned and innate behavior, because the two are always integrated, a separate consideration of learned behavior does give us additional insights into the processes controlling behavior and the background to deal with some behavioral problems.

We tend to think of learned behavior as responses we intentionally teach animals, such as parlor tricks or obedience tasks, or we may recognize learned responses in undesirable behavior that animals pick up on their own, such as begging for food or running away. To focus on these few instances of learning is inaccurate because learning actually affects an animal's behavior continuously. New responses are always being acquired and old ones lost on a day-to-day basis, usually without our knowledge.

Most authorities use the terms learning and conditioning interchangeably. The usual definition of learning is a process by which behavior is acquired or changed through reacting to a situation, provided that the acquired behavior cannot be explained on the basis of maturation or temporary states of the organism, such as reproductive status or fatigue.

In this book acquired behavior is dealt with in two chapters, the previous chapter on early experience and this one on learning. To a major extent the processes related to early experience involve the development of emotional reactions and attachment preferences for species, sex, food, habitat, and so forth. Animals may be excessively fearful, cautious, or aggressive as a function of prenatal stress or maternal deprivation. Classical conditioning, a discussion of which will constitute the first part of this chapter, can affect an animal's emotional behavior also, but the influences of classical conditioning on emotional behavior can occur in adulthood as well as in early life. The major thrust of this chapter deals with the nonemotional aspects of learning that are acquired on the basis of tangible rewards. Such behavior is referred to as operant (or instrumental) learning because, through some action of muscular movement, the animal's responses operate on the environment. Unlike the relatively permanent behavioral responses acquired on the basis of early experience, behavior resulting as a function of operant learning is acquired and lost quite easily.

Historically, there have been a number of theories or approaches to the study of learning. In all instances the principles of learning have conventionally been outlined as functioning independent of an animal's native response tendencies. The same principles can be applied to teaching rats to run a maze and teaching horses to count. In general, the principles work surprisingly well for such diverse tasks. However, species do differ in their innate tendencies or abilities to learn particular re-

sponses, and these differences are a reflection of evolved behavioral predispositions in the wild ancestors. In domesticating
animals humans have taken advantage of specific innate tendencies and built on behavior that is most easily shaped by learning. Hence, we teach dogs to hunt and retrieve, horses to carry
riders and race, and cows to line up and wait patiently outside
a milking parlor. We know that dogs will learn certain tasks for
praise or affection much more readily than cats, because dogs
are much more socially oriented toward people.

The approach to learning utilized in this chapter, while generally accepted as the most pragmatic, does not encompass all
the phenomena of acquired behavior. In the principles of operant learning it is assumed that responses are acquired because
there is some reward or reinforcement that often follows the
response. However, the phenomenon of imprinting, the development of attachments, and other aspects of early experience
discussed in previous chapters are not easily explained on the
basis of rewards only.

Learning by observation also does not necessarily involve
rewards. We know from personal experience that both people
and animals can learn by watching others and imitating them.
Relatively simple observational learning has been experimentally
documented in monkeys, dogs, cats, and rats. Typical of the
evidence of such learning is an experiment on rats where naive
rats were situated in a test chamber adjacent to an identical test
chamber containing thirsty rats that had already been trained
to press a lever for a water reward. The naive rats, who were
able to watch through a Plexiglas partition the trained rats working for their water rewards, also had a lever in their chambers
that they were free to press for a water reward. Most naive rats
soon started pressing their own levers and were able to obtain
the water reward. Meanwhile naive rats that could only see an
empty chamber next door learned lever pressing by trial and
error, but at a much slower rate. Naive rats who could watch
other naive rats in the adjacent chamber performed the poorest,
presumably because the other naive rats distracted them or misled the test rats by behavior that was not related to lever pressing
(Zentall and Levine, 1972) (Figure 8-1).

An example of the more complex type of evidence used to
demonstrate learning by observation in animals is an experiment

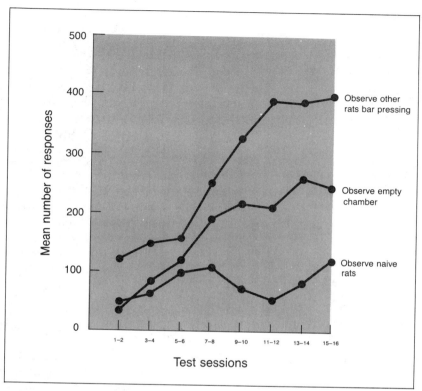

Figure 8-1 Evidence of learning by observation: Laboratory rats were allowed to observe other rats make bar-press responses for water rewards in a test chamber. Those that observed other rats bar press learned more rapidly than did rats that could only observe an empty chamber or naive rats in the adjacent chamber. (Modified from Zentall and Levine, 1972.)

in which puppies were taught to pull a food cart on a runner by means of a ribbon while other puppies watched from an adjacent compartment. Observational learning was demonstrated in the savings in time in the first trials when the observers were given the same problem to solve (Adler and Adler, 1977).

Observational learning could be adaptive in the wild because animals, particularly young ones, can learn a new response without having to undergo lengthy and hazardous periods of trial and error learning. For social animals observational learning allows a group to benefit from the experience of one or more of its members in terms of finding resources of food, water and safety (Crook and Goss-Custard, 1972).

■

PRINCIPLES OF CLASSICAL CONDITIONING AND HABITUATION

Any discussion of animal learning brings to mind the work of Ivan Pavlov, who, around the turn of the century, was the first to outline some laws of learning that appeared to hold for all animal species. The type of learning that Pavlov worked on is referred to as classical conditioning or conditioned reflex learning. This type of conditioning involves taking an innate response that is a reflex, or is reflexlike, and bringing it under the control of a neutral (conditioned) stimulus. The chief characteristic of classical conditioning is that only responses that are innate by nature can be conditioned. These are basically visceral responses, including changes in salivation, blood pressure, heart rate, milk ejection, and pupillary size. Spinal reflexes, such as the pain-withdrawal reflex, can also be conditioned.

The importance of understanding classical conditioning is that the visceral systems, especially those involving the cardiovascular, gastrointestinal, and pulmonary organs, which play a role in emotional behavior, can be conditioned by social and environmental stimuli. Visceral responses are "involuntary" and are very difficult to bring under the control of operant conditioning.

A partial list of visceral responses that can be conditioned is given in Table 8.1. The table lists the native, or unconditioned, stimuli that innately elicit the responses and also examples of common neutral stimuli (conditioned stimuli) to which these visceral responses are often conditioned.

The process of classical conditioning involves presenting a neutral stimulus immediately prior to the unconditioned stimulus. After a number of pairings, ranging from one to several depending on the response being conditioned, the neutral stimulus will eventually evoke the visceral response alone. The conditioning is the most efficient if the neutral stimulus precedes the unconditioned stimulus by an interval of a few seconds or less. Some types of conditioning will still occur, however, with a much longer time lag.

If the conditioned stimulus is repeatedly presented to the animal and the response elicited without giving the uncondi-

TABLE 8-1   List of visceral responses that are subject to classical conditioning

Natural response	Natural stimulus	A common conditioned stimulus
Salivation	Taste of food in mouth	Sight of food
Milk ejection	Massage of udder by calf	Banging of milk cans
Secretion of insulin	Ingestion of sugar	Smell of food
Emotional activation	Painful stimulation	Sight of person who administers pain
Nausea	Food poisoning	Taste of food
Asthmatic reaction	Foreign protein	Environment where foreign protein is administered

tioned stimulus, the conditioned stimulus eventually loses its strength to elicit the response. This process is called extinction.

An excellent example of conditioning of visceral responses is conditioned changes in blood glucose. The amount of sugar, or glucose, in the blood is regulated by an interplay of feedback controls. After a meal, when blood glucose concentration tends to rise, the pancreas secretes insulin, a hormone that enables most tissue to take up the glucose and burn it for energy or store it as fat. In the absence of sufficient insulin blood glucose levels remain high; this is a characteristic of diabetes mellitus. When blood glucose falls, other hormones are secreted, including glucagon from the pancreas and corticosteroids from the adrenal cortex, which cause the liver to convert protein into glucose or release glucose from stored glycogen. The secretion of these hormones is regulated directly or indirectly by the brain.

There are two natural responses involved here that might be subject to classical conditioning. One is the secretion of insulin, which lowers blood glucose, and the other is the secretion of the hormones that raise blood glucose. Like the study of many phenomena, this one was examined in the laboratory before the natural role of the response was fully appreciated. If an olfactory stimulus is paired repeatedly with an injection of an amount of insulin that causes a moderate lowering of glucose (which is followed by the secretion of glucagon and corticosteroids), then the olfactory stimulus alone can come to evoke the secretion of glucagon and corticosteroids, raising blood glucose without a previous fall in glucose. On the other hand, an olfactory stimulus repeatedly paired with an injection of glucose sufficient to cause an insulin response (and a lowering of blood glucose) can come to evoke the insulin response and induce a lowering of blood glucose (Woods and Kulkosky, 1976).

This type of conditioning appears to work in nature, since the taste or smell of foods that precedes the meal-related elevation of glucose can become conditioned and so cause the release of insulin before the absorption of food. When rigid feeding schedules are followed, secretion of insulin can even become conditioned to a time of day, so that blood glucose levels fall slightly then whether the animal is fed or not.

The conditoned visceral responses related to aversive stimuli are of great practical interest. Neutral stimuli paired with aversive stimuli can elicit the responses we associate with fear. These responses include an increase in blood pressure, sweating, slowing of intestinal motility, and general emotional activation. Some of these visceral reactions stem from the release of hormones from the adrenal medulla (epinephrine) and adrenal cortex (corticosteroids) and others are direct neural responses. Neutral stimuli can be conditioned to elicit both the hormonal and the neural responses. People who frequently punish animals with aversive stimuli can come to evoke all of these visceral responses without administering pain.

Pain is one type of aversive stimulation, but there are others. For many animals restraint is also considered aversive. A sudden change in stimulus intensity, such as falling or being thrown in the air, and very intensive stimuli, such as bright lights or loud noises, are also aversive. Consider the common practice of slapping a dog with a newspaper. One reason this might be considered aversive, in addition to all the emotional displays on the part of the person using a newspaper, is that the newspaper creates an intense sound.

There is evidence that some visceral responses to aversive stimuli can be conditioned by only one or two pairings with a neutral stimulus. A dog that gets shot with a gun once may acquire, from that one exposure, a severe emotional fear reaction to the sound of all gunshots and firecrackers, which previously did not disturb it. The value of electric shock as punishment stems from its ability to produce conditioned responses with one exposure.

In nature the emotional activation and escape responses produced by aversive stimulation have a good deal of survival value. On the domestic scene these fear responses are not particularly adaptable. If we are attempting to train or establish desirable behavioral patterns, the elicitation of visceral responses creates an emotional condition that may be incompatible with the behavior we are attempting to establish.

A learning process that also alters visceral and emotional responses is that of habituation. To understand habituation we have to recognize that animals have a number of innate fear reactions to loud noises, strange objects, and unusual sensations that are not in and of themselves harmful. This is quite adaptive in the wild where a built-in fear response to these stimuli often protects an animal from a harmful stimulus that may follow. The same responses are displayed on the domestic scene. Hence we see that horses are naturally upset by the pressure of a saddle on their backs because in the wild such a sensation would usually be caused by predators. When animals are repeatedly exposed to the stimuli which evoke the fear reactions, they become habituated to the stimuli and eventually lose the fear reactions.

In dealing with domestic animals we are often concerned with eliminating an innate emotional fear response that has not been habituated or extinguishing a response that has been acquired as a result of classical conditioning. The most successful way of eliminating these responses is by presenting the fear-inducing stimuli repeatedly but at low intensities that are gradually increased. This process is call systematic desensitization. A common example is in saddle-breaking a horse: the animal is gradually exposed to having something on its back.

The production of conditioned food aversions has become one of the more frequently studied phenomena because it appears to be one of the more practical examples of classical conditioning and because it differs somewhat from the conventional classical conditioning paradigm (Garcia et al., 1974). In Chapter 5 we have seen the role of this phenomenon in the control of feeding behavior and production of anorexia. If an animal is given a drug such as apomorphine or lithium chloride, it experiences a sudden brief bout of nausea and gastrointestinal illness from which it rapidly recovers. If given an arbitrary food such as milk, sweetened water, or meat before the onset of the gastrointestinal illness, the animal then treats the taste of the arbitrary food as poisonous, because the taste of the food has been conditioned to evoke feelings of nausea similar to those produced by the drug alone. The conditioning is usually established in one or two trials, rarely requiring more than three. A characteristic of conditioned aversions in mammals is that the conditioned stimuli are virtually always taste cues and not visual or auditory cues that accompany the taste. In avian species, in which the chemical senses are very limited, similar conditioning is associated with visual rather than taste cues (Bolles, 1973). A major characteristic of conditioned taste aversions is that the aversions are maintained wherever the tastes are encountered, even in novel places. The adaptive value of such taste aversions is easy to see. For herbivores, bitter-tasting plants often have alkaloid toxins, and it behooves an animal to capitalize on just one experience with a toxin regardless of location.

Another interesting example of conditioned visceral responses is experimentally induced asthma. By making guinea pigs repeatedly asthmatic with the injection of an allergen in a certain test chamber, it has been found that simply putting them in the test chamber alone later induces the full asthmatic reaction (Ottenberg et al., 1958).

■

PRINCIPLES OF OPERANT LEARNING: ANIMAL TRAINING

The concept of operant learning is simple: If a response is followed by a reward (reinforcer), the probability of the response occurring again increases (Figure 8-2). When barking is followed by a reinforcement such as a chunk of meat, it tends to be repeated. However, barking will be reinforced only when a person is around to deliver meat chunks, and so we can speak of a stimulus, in this case a person holding meat chunks, that evokes the response of barking. Since the dog may bark as frequently as it pleases, or not at all, it is said that the response is emitted rather than elicited.

This rather simple approach to learning is based on the principles originally formulated by B. F. Skinner (1938) and modified by learning specialists since then. The concepts are the most widely used and accepted approaches to operant learning.

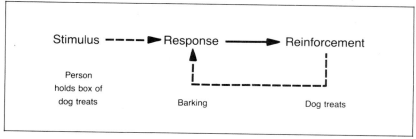

Figure 8-2 Paradigm of operant learning for barking: When a response is followed by a reward or "reinforcement," the response tends to be repeated. The stimulus term refers to the stimuli present when responding will be reinforced. Thus an animal learns to respond mostly when the stimulus is present. However, since the response is not forced (i.e., it is voluntary), the line between "stimulus" and "response" is broken.

The Operant Level

Behaviorists speak of the operant levels of various behavioral patterns. The operant level is the frequency of a behavior that exists before learning of the task. Barking, for example, is natural for dogs and occurs spontaneously at a certain rate. To teach barking we can simply wait for a dog to bark and then present it with meat each time barking occurs. Eventually the frequency of barking will increase in our presence. We can also use special procedures to increase the operant level of barking, such as making barking sounds ourselves. If this evokes barking in the dog, the process of teaching it to bark for food is hastened.

One example of how a response that exists at the operant level is learned is illustrated by the development of begging behavior in dogs. Consider a dog that is lying under the family table at dinner. The dog smells meat, is aroused, gets up, and begins to walk around. Let us assume that the animal has never been reinforced for begging in the past. As the dog paces around the table, it may eventually rest its head on the leg of one of the children. This head-resting behavior previously has been a way of soliciting petting from the child. The child may then give the dog a scrap of meat from its plate while the dog is head-resting. The dog eats the meat, walks around a bit more, comes back to the child, and engages in head-resting again. The child again gives the dog some meat. It will not require many more meat reinforcers before the behavior of head-resting is firmly established and the dog has learned a type of begging behavior. In this example, the dinner table, the smell of food, and the presence of the child at the table are all stimuli that evoke the head-resting behavior.

Extinction

After a response has been learned it is maintained as long as reinforcement is at least occasionally presented. If reinforcement is permanently withheld, animals will make fewer and fewer responses. Eventually the rate of responding will drop to the previous operant level. This is called extinction. Extinction is an active process, requiring responses to be made, and should be distinguished from forgetting, which is a passive process representing a decrement in response strength that occurs simply with the passage of time.

Types of Reinforcement

Some of the things that constitute the class of reinforcers, such as food, water, sexual activity, and exploratory behavior, require that the animal be deprived of the stimuli before they can be reinforcing.

Although we tend to feel that food can only be reinforcing if an animal is hungry, this is not entirely true for dogs. Favored foods such as hamburger may be highly reinforcing to animals that have already consumed a large meal.

There are reinforcing stimuli that clearly do not require prior deprivation. Social contact and petting are examples for dogs even if they are already receiving a great deal of attention (Hart, 1979).

Warm air or radiant heat is obviously reinforcing to an animal that is cold. By the same token cold air is reinforcing to an animal in an environment that is too warm. Experiments have shown that rats will press a lever to turn on a heat lamp when they are cold and turn on a fan when they are too hot. A refinement of this procedure was used by Baldwin and Ingram (1967) to examine the way pigs would control their environmental temperature by learned responses. Pigs placed in a cold environment were allowed to press a switch to turn on heat lamps. The response was learned simply by trial and error. Some pigs learned the task in less than 30 minutes (one pig in less than five minutes) and some took over seven hours. Once the re-

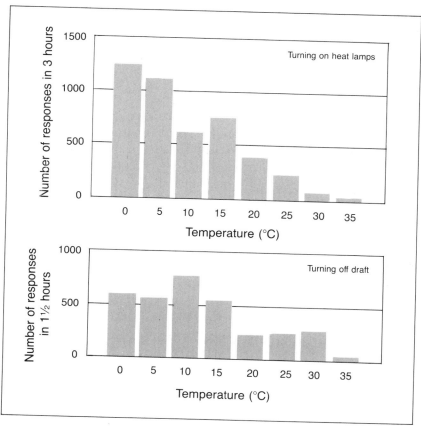

Figure 8-3 Pigs can learn to turn on a heat lamp or turn off a fan when they are cold. The top bar graph, from data on one pig, shows that as ambient temperature increased, the pig's tendency to turn on the heat lamp declined. Each press on a switch activated the heat lamp for three seconds. The bottom bar graph shows data for a pig that learned to turn off a fan creating a draft. Each push of a switch turned off the draft for five seconds. As the ambient temperature became warmer, the pig's tendency to turn off the draft declined. (Data drawn from Baldwin and Ingram, 1967.)

sponse was learned, the pigs were placed in the test chambers for 30 to 60 minute periods; the temperature in the test chamber ranged from below freezing to 40°C. A single press on the heat lamp switch would turn on the heat lamps for three seconds. As one might expect, the pigs frequently turned on the heat lamps in the colder environment and left the heat lamps off in the warmer environment (Figure 8-3). In another experiment the test chamber was fitted with a fan that ran continuously except for a five-second period after the turn-on switch was pressed. Pigs learned to turn off the fan, which created a draft, when the environment was cold, but they tended to let the fan run when the chamber was warm (Figure 8-3).

Experiments such as these on pigs, in which the animal is allowed to make adjustments in its environment through learned responses, give us information about an animal's comfort level. We can then design animal housing on the basis of such animal-oriented information rather than our own guesses. It could well be that an animal's comfort range is lower than what we believe, and we may save money by heating less.

Work by Baldwin and Meese (1977), again on pigs, indicates that there is a quantitative as well as a qualitative aspect in determining an animal's preferred environment. Pigs were placed in a test chamber where they could turn lights on and off by sticking their snouts into a photoelectric beam. Interruption of one photoelectric beam turned the lights on and of the other turned the lights off. Over a period of several days the pigs kept the lights on 72 percent of the time, but they turned the lights off for a short period at least once per hour.

When one reconsiders the range of species-typical behavioral patterns discussed in the preceding chapters, several natural reinforcers of behaviors come to mind that are related to an animal's physiological state. Sleep is obviously reinforcing to a sleepy animal. Finding an innately appealing place to eliminate is reinforcing to an animal with a full urinary bladder or rectum. Finding newborn that have strayed from the nest is reinforcing to a mother.

Animals will perform learned tasks or learn new tasks to obtain such reinforcers. For an expectant mother rat, acquiring appropriate nest material can be reinforcing, but only when she nears the physiological state when it is natural for her to build a nest. This was illustrated in an experiment in which pregnant female rats were placed in a cage where they could obtain nesting material only by pressing a lever, which delivered narrow, four-inch-long strips of paper. The female rats quickly learned to work the lever press for a strip of paper (Oley and Slotnick, 1970). Prior to parturition they worked for only a small amount of paper, sufficient to build sleeping nests, or no nests at all if it was warm. On the day of parturition huge nests were built out of strips the rats got by lever pressing (Figure 8-4).

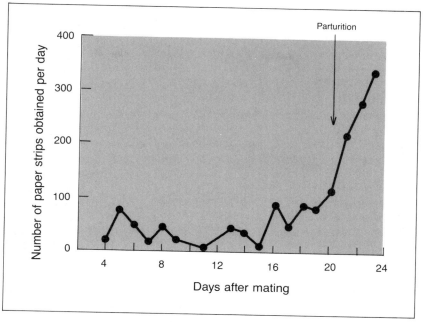

Figure 8-4 Rats will learn to lever press for paper strips with which to build a nest. However, the performance of a female rat approaching parturition exceeds her performance at other times. (From Oley and Slotnick, 1970.)

Reinforcers are considered to be positive when their presentation increases the probability of the responses preceding them occurring again. Hence food, sexual activity, and water are positive reinforcers. Negative reinforcers are aversive stimuli that, when removed, increase the probability of the responses immediately preceding them occurring again. Animals can learn to diminish them by taking some action. Fear, for example, is aversive and the reduction of fear is reinforcing. Animals will learn to escape or to act aggressive if such behavior reduces fear.

Negative reinforcement is easily confused with punishment. Punishment involves the presentation of aversive stimuli such as pain, intense noises, or social isolation in response to an undesirable behavior. Although punishment may interrupt the behavior, it does not extinguish it. In negative reinforcement aversive stimuli are removed when the animal makes the response.

The best example of negative reinforcement is the way fear enhances aggressive behavior. Fear is an unpleasant or aversive state, and when threatening, snapping, or kicking drives away someone who causes fear, the aggressive responses are reinforced.

Things that are biologically reinforcing are referred to as primary reinforcers. These include food, water, sexual activity, oxygen, warmth, sweet substances, exploratory activity, and social contact. Much of an animal's learning is reinforced by secondary reinforcers. These are stimuli that were previously neutral but that, after being paired with primary reinforcers, have taken on reinforcing properties. A box containing dog biscuits takes on the reinforcing properties of the biscuits because it is always paired with the biscuits. Sounds associated with primary reinforcers may also take on reinforcing properties, as revealed by an advertisement in the lost-and-found column of a local newspaper: "Lost, female calico cat that answers to the sound of an electric can opener."

Once secondary reinforcers cease to be paired with primary reinforcers (at least periodically), they lose their reinforcing properties. Commonly used verbal secondary reinforcers are the words "good boy," or "good dog." Dogs can be taught to sit, stay, or heel with these secondary reinforcers. However, if the words are never followed by affection, food, or other primary reinforcers, they eventually lose their reinforcing properties. The electric can opener would lose its properties in maintaining a cat's behavior of running to the kitchen if its sound was no longer paired with the presentation of food to the cat.

Stimulus Control

Certain stimulus configurations indicate to an animal that a particular response will be reinforced. Thus, the response of rolling over for a dog will usually only be reinforced in the presence of a person who says "roll over." Once the phrase "roll over" evokes the response, it is referred to as a discriminative stimulus.

Even after an animal has learned to roll over in response to the words "roll over," there is still a range of physical characteristics of the stimulus to which the response may be made. A dog may roll over regardless of what person gives the command even though the initial response was conditioned by just one person. This is an example of stimulus generalization. The dog has generalized from the particular voice of one person to a range of voices of many people. A dog that sits up and begs when its master takes a box of dog biscuits off the shelf may also sit up and beg when a box of saltine crackers is taken off the shelf because of the relative similarity between the two boxes. This too is generalization.

It is possible to narrow down the aspects of the stimulus to which animals may respond. If only one person ever reinforces a dog for performing the trick of rolling over, while other people give the command but never reinforce the behavior, the animal's behavior of rolling over will extinguish for all voices except the voice of the person who continues the reinforcement.

Animals can learn a large number of discriminations with no intentional help from people. They may react to the smell of certain people. Detecting the time of day is another type of discrimination. Horses and cattle may wait in certain areas for food when their waiting behavior, at that time of day, has been consistently followed by food reinforcement.

The claims of unusual reasoning powers in animals—the ability to count, add numbers, tell time, and so forth—are a reflection of the ability of animals to detect subtle changes in facial expressions or movements of which their owners may not be aware.

The classic example of an animal's being able to discern very subtle cues and respond appropriately to them is the story of Clever Hans (Figure 8-5) (Rosenthal, 1965). The remarkable intellectual feats of Hans became headlines in Berlin around 1900. Hans was owned by an old schoolmaster named von Osten. Herr von Osten decided to tutor Hans as he would a schoolchild and patiently worked with the horse for over four years. Since Hans could not speak, he had to respond to von Osten's questions by shaking his head, nodding, and especially by tapping his right forefoot. Herr von Osten had taught Hans to tell time, recognize some music scores, name coins, and perform arithmetic. He could even do fractions by counting out separately the numerator and the denominator.

It was not long before people came from all over to see Hans perform. Circus trainers, zoo directors, explorers, psychologists, and philosophers saw Hans and claimed he was an equine genius. As Hans became more famous he was represented as a toy, in pictures, and on liquor labels. Trickery was suspected but could not be proved because even when von Osten was asked to leave and another questioner took over, Hans performed cor-

Figure 8-5 Clever Hans was a horse that had experts fooled into believing that it correctly solved problems in multiplication and division by tapping its right forefoot. (Photograph from Rosenthal, 1965.)

rectly. Eventually a well-known authority, Professor Stumpf, and his young colleague Oskar Pfungst became interested in trying to understand Han's remarkable abilities. They suspected something when they learned Hans's performance fell off at dusk or if he wore blinders. Finally one experiment proved crucial in understanding the secret of Hans's intellectual abilities. Two questioners each whispered a part of an arithmetic problem to Hans. He might be asked to add three plus five, for example. If the questioners agreed in advance about the problem to be given to Hans, he performed perfectly, but if they did not tell each other the numbers they were asking, Hans failed miserably. It was soon evident that Hans was cueing onto very subtle facial or bodily movements of his questioners. When Hans had tapped out a certain number of responses, there was usually some subtle sign of relief on the part of the questioners that Hans took as a cue to stop tapping.

Shaping

The concept of shaping is important in understanding how responses that do not exist at the operant level can be learned.

Figure 8-6   Cats can be taught proper toilet training utilizing the techniques of shaping. (From Hart, 1975. Reproduced with permission of Veterinary Publishing Co.)

A simple illustration of such shaping is when we attempt to teach dogs and cats to use pet doors. One of the types of cat doors commercially available is constructed in such a way that the cat has to learn to flip up the door from the outside. This keeps out feline intruders who do not know the trick (the door design also keeps out the rain). The cat owner must teach the resident cat access from the outside. A string can be attached to the door to hold the door open until the cat is accustomed to coming and going freely. The door is gradually lowered over a period of several days, and eventually the cat is having to crouch down and push the door up when it comes inside. In the final phase it must use its paw to push the door up.

The use of shaping can also be important in teaching proper toilet training to cats. The procedure used to train the cat in Figure 8-6 to defecate and urinate in the toilet was rather a straightforward example of shaping (Hart, 1975). This training takes advantage of the cat's natural eliminative tendencies; the reinforcement is being able to eliminate.

To start the learning process the litter box is placed in the bathroom, and after the cat is regularly using the litter in this location, a litter tray is fashioned out of the toilet seat by covering the bottom of the seat with a sheet of clear plastic. A cardboard rim can be cut in the shape of the toilet seat to reinforce the plastic on the bottom. With plenty of litter material in the toilet seat, a cat should readily use the new litter tray. Cats will learn to stand on the edge of the toilet seat because of the uneasy nature of the center. Over the next few days the litter should gradually be removed, until finally all the litter and the plastic sheet have been removed.

Shaping of more complex behavior involves starting with a response that is already in the animal's repertoire and working toward a desired behavioral goal. Reinforcement is given only to those responses that are in the direction of the desired behavior. Reinforcement is withheld from those responses that are not in the direction of the desired goal. As training continues it is necessary to keep changing the criteria of acceptable responses, demanding a closer approximation to the desired goal behavior before reinforcement is given.

In shaping advantage is often taken of the variability that occurs in all animal responses to a given command. For example, teaching a dog to roll over by starting with the response of lying down takes advantage of the variability in the act of lying down in that sometimes a dog lies more flat than at other times. When the dog is lying down, one may give the command "roll over" but reinforce lying down only when the animal is completely on its side. The next step is to keep repeating the phrase "roll over" but withhold reinforcement until the animal is lying flat on its side and raising one of its paws off the floor. The criterion for reinforcement is thus changed from lying down flat to lying flat with one or more paws raised. Once the dog begins to lift its legs, one can observe variability in the degree to which the legs are raised and demand a progressively greater degree of leg raising. Through several progressive steps of raising the criterion for reinforcement, the dog should eventually be raising its legs almost vertically. From this point on it is only a short step to raising the criteria once more to where the animal must tip itself over, thereby automatically completing the response of rolling over.

In shaping a response such as rolling over it is necessary to carefully note very small changes in the direction of a response. Sometimes it is necessary to use a reference point to pick up small changes. It is also necessary to signal the animal within a fraction of a second when its response is correct. This is often done by the use of secondary reinforcers. A dime store cricket

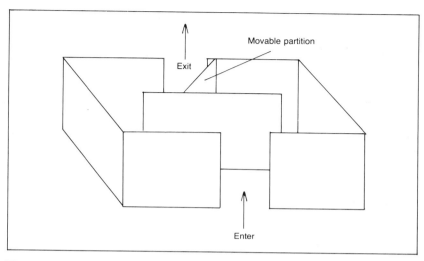

Figure 8-7 This maze for horses was positioned where the horses usually left the barn. The reinforcement for running the maze was to be released to outside pens. This type of learned task has a built-in continuous reinforcement. (Maze design from Kratzer et al., 1977.)

or whistle that has been paired previously with food can be sounded just as the animal is lifting its legs off the floor. At the sound of the secondary reinforcer the dog will rush over to get some food. Another trial is initiated by giving the command "roll over."

Reinforcement Schedules

The commonest reinforcement procedure when teaching an animal a response is to give it a reinforcement immediately after the desired response almost every time the desired response is made. This is referred to as continuous reinforcement. Animals learn rapidly with such a schedule. With the shaping procedures, it is imperative that continuous reinforcement be employed.

From the practical standpoint some types of learned behavior must be reinforced on a continuous basis because of the nature of the task. Consider teaching a horse to run a maze such as that shown in Figure 8-7. This maze was positioned in the door of a barn so that when the horses wanted to leave the barn to obtain water and be with other horses, they had to negotiate the maze (Kratzer et al., 1977). Every time the maze was successfully run, the horse was reinforced. Horses learn such a maze fairly readily. This type of response takes longer than lever pressing, barking, pawing, and so forth, and it is not feasible to require the horse to negotiate the maze more than once per reinforcement.

It is obvious that with responses that can be made rapidly a response need not be reinforced every time it occurs in order for the learned behavior to be maintained. In most situations learned behavior is reinforced intermittently. Consider the behavior of a dog scratching at the door to be let in. The dog is reinforced intermittently because it scratches 20 to 30 times before it is allowed in.

The most important consideration involved in comparing intermittent reinforcement with that of continuous reinforcement is that learned behavior is much more resistant to extinction when it is maintained on an intermittent schedule than when it is reinforced on a continuous basis. Another way of putting it is, the number of nonreinforced responses that will occur before a response is extinguished is much greater in the case of intermittent than continuous reinforcement.

With the increased resistance to extinction and the prolonged and variable responding that follows intermittent reinforcement, behavior may be shaped to an intense form. This is especially true of objectionable behavior that is not intentionally conditioned. For example, sometimes people reinforce intense door scratching by dogs because it is only when the dog exhibits prolonged and intense scratching that they let it in.

There are two types of intermittent reinforcement schedules. One of these is a ratio schedule. If a person is delivering reinforcement and decides that an animal must make an exact number of responses before reinforcement, this is using a fixed ratio schedule. In practice most reinforcement given on a ratio basis involves variable ratios: an average number of responses is required per reinforcement. Under either fixed or variable ratio schedules learned behavior is performed rapidly since the sooner the required number of responses is made, the sooner reinforcement is obtained.

The other type of reinforcement schedule is the interval schedule. This may also be fixed or variable. In a fixed interval schedule an animal is not reinforced for another response until a certain period of time has elapsed since the preceding reinforcement. This type of schedule is involved when livestock are seen waiting for food at the same time and place each day. A

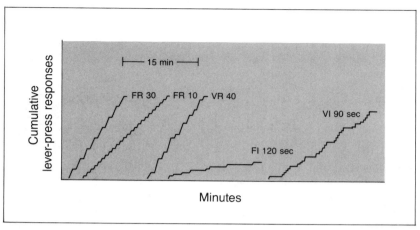

Figure 8-8 Examples of differences in rate of responding as a function of reinforcement schedule: Cows were trained to press a plastic plate for a grain reward. Rapid responding was obtained with fixed ratios of 30 and 10 (FR 30 and FR 10) and a variable ratio of 40 (VR 40). A fixed interval of 120 seconds (FI 120 sec) resulted in slow responding, but a variable interval of 90 seconds (VI 90 sec) yielded higher response rates. (From Moore et al., 1975.)

dog belonging to a child may have learned to wait by the window in the afternoon in anticipation of the child's return from school. If the child returns home at four o'clock each day, a fixed interval schedule is involved. In this situation the dog would probably begin to display waiting behavior quite close to four o'clock. If the child returns home at a different time each day, the dog is reinforced on a variable interval schedule, and it is likely to display waiting behavior rather frequently throughout the day.

All four reinforcement schedules might be used under experimental conditions, especially those in which animals are free to press a lever or move some device rapidly. An animal's pattern of responding reflects the type of reinforcement schedule. This is evident in Figure 8-8, which shows the response pattern typical of cows that had been trained to press a plastic plate with their chins to obtain a grain concentrate reward (Moore et al., 1975). A fixed or variable ratio schedule produces the most rapid responding, a long-delay fixed interval the slowest rate of responding, and a variable interval, a steady rate of responding somewhat in between that of ratio and fixed interval schedules. Such response characteristics by animals generally approximate the mathematically calculated response pattern that would match least effort against maximum payoff.

∎

COMPARATIVE ASPECTS OF ANIMAL LEARNING

Although there are many similarities among species in the way the principles of both operant learning and classical conditioning can be applied to understanding behavior, there are some obvious species differences in the way animals can be trained. Teaching a dog obedience tasks, or even parlor tricks, is a different matter than teaching a cat the same responses. Horses can be trained to run more easily than cattle can. Many of the differences in learning behavior reflect innate differences in the wild ancestors resulting from a long evolutionary history of adaptation to ecological and social environments. In addition to differences in sensory-motor capacity among animals, we can examine how brain complexity and evolutionary development relate to learning ability.

Preparedness to Learn

Different species are prepared to make some learned associations and are unprepared, or contraprepared, to make other types of associations (Seligman, 1970; Bolles, 1973). It is almost impossible to teach a dog to yawn for a food reward although it is relatively eash to teach it to walk on its back legs. Cats have been taught to perform a variety of tasks to escape from a puzzle box, including pulling strings, tripping levers, and pushing buttons, but it is practically impossible to teach a cat to lick or scratch itself to escape (Thorndike, 1964).

Some of the most interesting examples of how an animal's natural response tendencies can influence its learning ability come from the work of the Brelands in their efforts to train a variety of wild and domestic animals for department store displays, television commercials, and state fairs (Breland and Breland, 1961, 1966).

In one of their trained animal acts a Bantam chicken emerges from a restraining compartment, walks about a meter onto a raised platform, and pulls a loop hanging from a box, which starts a four-note tune. Standing on the platform closes a microswitch, which starts a timer. The chicken then scratches vigorously for 15 seconds until an automatic feeder fires in the restraining compartment, after which the chicken goes back into the compartment to eat, and the door automatically closes.

The popular interpretation of this act is that the chicken turns on a jukebox and dances. The original intent was to have the chicken just stand for 15 seconds and listen to the music. However, the chickens almost always tended to scratch while they were waiting for the automatic feeder to fire, and so the trainers decided to let this be part of the act. It was noted that in the course of three hours, chickens could turn out about 10,000 unnecessary scratching movements.

The Brelands attempted to train some raccoons to pick up coins and put them in a box (the box was eventually going to be painted to look like a coin bank). The raccoons were started with just one coin and had no problem in picking it up. But they had a great deal of trouble in dropping it into the box. They would start to drop it, rub it against the side of the box, pull it out, push it in, and so forth. With two coins the raccoons really had problems. They continuously rubbed them together, dipping into the container and pulling them out again. This behavior is, of course, reminiscent of the raccoons' tendencies to handle and examine their food in water (often referred to as "washing" the food).

The trainers wanted to teach pigs to pick up large wooden coins and deposit them in a piggy bank. The pigs conditioned quite rapidly and seemed to have very little trouble in responding to interval reinforcement ratios. In the course of training it was found that initially pigs would eagerly pick up a coin, carry it to the bank, get another coin, carry it rapidly, and so forth until they had dropped several coins in the bank. Over a period of weeks, however, the carrying behavior became slower and slower, and instead of simply carrying the wooden coins, the pigs would repeatedly drop them, root them, drop them, root them, then toss the coins in the air, drop them, and root them some more. This drift toward species-specific behavioral tendencies was intensified by making the pigs more hungry.

One of the contracts of the Brelands called for attempts to train a cow for a movie script in which the cow was supposed to approach a miner's campfire, kick over his bucket, chase him around the campfire, knock down his tent, and display a wild bullfight with the miner. The trainers could condition every-

thing but the kicking, and so they simply used shaping to condition the cow to chase the miner around, knock down the tent,
and engage in a fight for food reinforcement. However, all of
this was done at a ridiculously slow, bovine pace. The reactivity
of the cow simply reflected her natural life style. Because a cow's
foodstuff is so low in nutritive value, she can only afford to spend
a low level of energy and time in obtaining food. The above
examples demonstrate that many aspects of learning are related
to an animal's innate predisposition, which puts constraints on
learning.

In addition to the genetic attributes that place negative constraints on what an animal can learn, we might also think of
situations in which particular learned responses might be po-
tentiated by genetic attributes. The acquisition of aversions to
the taste of particular foods, stemming from a single poisoning
and gastrointestinal illness associated with each of the foods, is
the best example. The learned aversion occurs in one or two
trials, is long-lasting, and is associated with a food flavor rather
than the visual or auditory cues that may also accompany the
illness.

A particular potentiation of learning also occurs with observational learning. The term observational learning is misleading in that it implies that all an animal has to do is watch
another one performing the accomplished task and then to im-
itate what it sees. This is a lot to ask of a member of the human
species (think of teaching a child to tie shoelaces, or teaching
an adult friend to ski), let alone animals. But if the animal's
learning is measured by the decrease in errors or improvement
in time to learn a new response, then the facilitative aspects of
observation are clear. The more familiar the animal is with the
movements involved, the more closely the animal watches the
demonstrator's goal-directed behavior, and the more the observer sees the demonstrator eliminate errors in performing, the
more rapidly will it learn the response on its own (Herbert and
Harsh, 1944).

Learning by imitation can be a highly adaptive trait in young
animals. A young animal must locate and consume food, avoid
predators, locate mates, and so forth. Although the responses
appropriate to the attainment of these goals can be acquired by
trial-and-error learning, the costs of errors in learning time could
be fatal. A mother can facilitate the learning in her young by
acting as a demonstrator. Thus, we have in the animal world an

example of a social transmission of information (Galef, 1975). If one takes a colony of wild rats and trains them to avoid a certain palatable food by lacing the food with a sublethal dose of poison, thereby inducing the adult rats to eat a less palatable diet, the young of the colony will only eat the less palatable food also even though they experienced no poisoning. When the young are separated from the adults, they still continue to eat the less palatable food.

How is this social transmission of acquired eating behavior accomplished? Galef mentions that the young usually take their first solid food in the immediate vicinity of a feeding adult. Also, adults tend to deposit large amounts of urine and feces near feeding sites, and young are more willing to take their first meals in areas marked in this way. After their first experiences with a food, rats tend to stay with the same food as a function of their inherent tendency to avoid novel diets.

Animal Intelligence and Brain Evolution

There is a limited amount of research suggesting some fundamental phyletic differences in animal learning that relate to brain evolution. The main spokesperson for this line of research is M. E. Bitterman (1965, 1975). He and his colleagues have studied mostly fish, turtles, laboratory rats, pigeons, and monkeys. An example of the types of processes examined by Bitterman and co-workers is resistance to extinction as a function of whether reinforcement is intermittent or continuous. We have seen that in mammals learned behavior is much more resistant to extinction when reinforced intermittently than when reinforced continuously. However, in the fish studied by Bitterman the learned performance of striking a target disk for food was more resistant to extinction after continuous reinforcement than intermittent reinforcement.

Another example of a basic difference between the learning abilities of different species of animals relates to the depressor effect on learned behavior. Depression of responding occurs in mammals when a reinforcement is shifted from a large reward to a small one, or from a highly favored reward to a less favored one. Thus, rats show a decrement in maze performance when the reinforcement is shifted from a favored bran mash to less favored sunflower seeds. Bitterman found that target-striking performance by goldfish did not show the depressor effect when the reinforcement was shifted from 40 Tubifex worms to four worms. The depressor effect also failed to appear with maze running in turtles.

Animals vary greatly in the relative size of their brains and the development of the cerebral cortex. Domestic mammals have a much more highly developed cerebral cortex than rats, for example. Yet it is not easy to demonstrate major differences in learning ability between rats and carnivores. There appear to be no special differences among mammals in the ease with which they acquire classical conditioning or simple operant learning such as maze running.

Typical of the types of learning problems psychologists have devised in an attempt to separate the intelligent from the not-so-intelligent species is something called conditional discrimination. In a standard discrimination task stimuli presented to animals might be wood objects of various shapes. When the "correct" one is touched, a food reward is delivered. In one two-choice conditional discrimination problem the stimuli could be a square and a triangle. When these forms are both large, choice of the square is correct, and when they are both small, choice of the triangle is correct. Rhesus monkeys have been shown to master this task easily, but most cats have a lot of trouble with it. Yet a few cats catch on as fast as monkeys (Warren, 1965).

Roughly the same results are obtained with oddity problems, in which the animal is presented with three objects, two of which are always identical and one of which always differs in shape or size. The animal is required to choose the odd object. Monkeys learn this task quite easily and can choose an odd object from three objects never seen before. Most cats learn to solve a single-oddity problem but fail to generalize the oddity principle to a novel set of three stimuli. The occasional cat, however, is able to generalize the principle as well as monkeys (Warren, 1965).

We can see, therefore, that differences between mammals are more quantitative than qualitative. There are huge individual differences among members of a species that may bridge differences between species. By not testing animals long enough or by testing too few animals, we might erroneously reach a conclusion that a certain task is impossible for animals of one species to solve. Negative findings—that is, findings that an animal cannot learn a task—are always subject to revision.

Delayed response is a complex task that one might expect many animals to fail. In a typical delayed response test one of three boxes in a room contains a food reward. The animal is shown which container has the food, the food is then covered, and the animal removed for a period of time before it is allowed to go after the correct container. The task involves the animal's "remembering" which container has the food for, say five to 30 minutes. Monkeys and cats are good at this task, but one might expect animals that are less visually oriented to perform less well. Initial work on goats suggested they could not perform this task. However, a more painstaking approach showed differently. If during the demonstration trial the goat's attention was directed to the correct location by the use of flashing signals, rattling the food bowl, and allowing them to eat briefly from the bowl, they performed beautifully. Even with a delay of as long as 24 hours goats still performed better than at the rate of chance (Soltysik and Baldwin, 1972).

Reversal learning is considered a complex task in which an animal learns to learn. A typical experiment is to teach an animal to run a Y maze, where going to the left or the right is the correct response. After an animal meets the criterion of almost perfect performance, the previous correct side of the maze is then made the wrong side. This process is repeated. Of course, it takes one error to know that the previously correct side is now the wrong side. Each time the side is switched one expects to find animals improving in the rate at which they reach a perfect performance. This is learning to learn, and some animals eventually learn to switch over to the other side after making just the one error.

Experiments reveal that almost all mammals tested can master reversal learning. What is perhaps the most interesting is the wide range of individual differences within a species. Fiske and Potter (1979) tested horses on such a Y maze and found that one horse performed almost flawlessly in switching from side to side, whereas another horse never reached perfection at the reversal learning. Interestingly, a subjective evaluation of the trainability of the horses by the trainer who had halter broken the horses was roughly correlated with their learning performance, but the relation was far from perfect.

Another learning task that reveals some interesting individual differences within a species is discrimination learning. Mader and Price (1980) taught Quarter Horses to perform a pattern discrimination by confronting the horses with an apparatus that held three small swinging doors. Mounted on one door was a checkerboard pattern and mounted on the two others was an entirely white board. When the horses pressed the checker pattern (regardless of position), the door swung open and allowed the horse to get some grain. Horses learned this task quite readily. However, there were major individual differences in the number of trials to reach an experimental criterion. There was no correlation between dominance of the horses and performance or between emotionality rating and performance. But Quarter Horses outperformed Thoroughbreds. Mader and Price mention that factors such as motivation for food or undetected differences in emotionality might explain the breed difference. They also offer an explanation that is in keeping with the idea of genetically acquired constraints on learning. Quarter Horses have been bred for a variety of tasks involved in working cattle that require rapid discrimination learning, whereas Thoroughbreds have been selected for success on the racetrack. The latter is a singular behavioral trait that does not rely much on learning ability.

Of interest from the standpoint of comparative animal intelligence is the series of studies by Scott and Fuller (1965) on problem-solving ability in different breeds of dogs discussed in Chapter 6. In the individual problem-solving tests, which included delayed-response, detour and manipulation tasks, there were clear breed differences, but no single breed performed best in all of the tasks. There was no evidence of all-around, superior problem-solving ability or "intelligence" in any breed.

Can we rank the domestic animals on the basis of intelligence? The answer depends on whether we can accept performance on one test as an indication of intelligence. Clearly the more we learn about the subtleties of testing and the interaction of specific sensory and motor abilities with learning, the more we recognize that it will probably be impossible to rank our common domestic animals in any intelligence hierarchy.

■

APPLIED AND PRACTICAL CONSIDERATIONS

Imagine the modern livestock farm in which pigs press levers to turn on heat lamps during cold weather or to give themselves a water spray cool-down in summer. Or think of the dairy farm in which cows are called in for milking from a pasture by a horn. Better yet, how about a portable alarm device activated by sheep when a predator arrives on the scene? Training animals to perform useful tasks to save a farmer time, effort, and money and to make life more comfortable for the animals is a logical application of the principles of classical and operant conditioning. With the companion animals, dogs, cats, and horses, we are more concerned with treating problem behavior that interferes with their role as our pets or riding companions. In many instances we can utilize the principles of classical and operant conditioning to resolve the problem to the animal's and our mutual benefit.

Learning and Training on the Farm

The principles of operant conditioning are useful in examining an animal's preference in temperature, humidity, light, contact with conspecifics, and other aspects of the environment. This is important because we maintain such a tight control over the environment of animals. Farm animals usually have little choice in environmental regulation. With behavioral methods we can learn something about what animals like. It is almost economically foolish not to take advantage of operant conditioning techniques to preclude some of the more onerous tasks around the farm. Consider the dairy farmer who is in the habit of going to a field and fetching the cows twice a day for milking. For the cows, it is reinforcing to have milk pressure in the udder released through milking, and if they are fed a palatable grain supplement during milking, movement toward the milking parlor is all the more reinforcing. Since the cows are already conditioned to start walking toward the barn at the sight or sound of the farmer, it is an easy task to condition them to proceed toward the barn with an inanimate auditory stimulus. In the case of a dairy farm in England, where an electric fence was already in use, Kiley-Worthington and Savage (1978) designed an alarm signal that could be attached to the electric fence. A modified automobile horn was used as the alarm. Timers operated the alarm at a designated time in the evening and switched off the electric fence. For the first week the farmer appeared in the usual way and coupled his calling with the alarm sound. After only three trials, 28 percent of the cattle began to walk toward the gate without seeing the farmer; by the end of one week 85 percent of the cows responded within five minutes and no cows remained in the field for more than 19 minutes after the alarm. This technique would be particularly useful on large farms where cows have to walk a considerable distance to reach the milking parlor. In the examples above, the investigators estimate that about one hour per day of the farmer's time was saved.

In addition to using conditioning techiniques as laborsaving schemes on the farm and as a method to investigate the preferences of livestock, tests of the learning ability of animals might be used to examine differences among individual animals. One thing that problem-solving and discrimination tasks have taught us is that there are pronounced differences among animals of the same species in ability to catch on to specific tasks. Some specific tasks might be directly related to what is desired of an animal as an adult. This could, for example, be important in horses. A passive test, such as a maze, that a foal must negotiate daily to get to its mother or other horses might constitute an appropriate test. If such a learning task proved to be a reliable predictor of performance as an adult, the extra effort and expense involved in behavioral testing would be economically justified. In addition, the progeny testing of foals for a behavior that predicts success as a working horse later cuts down on the time lag in deciding which of the breeding stock are providing the greatest return in value of their offspring.

Conditioning Principles and Behavioral Therapy

The use of learning principles to correct problem behavior is a whole field in itself. This approach to behavior is usually referred to as behavior modification. The principles touched on in this chapter can be used to understand how some behavioral problems develop (Hart and Hart, 1985). Some classic examples are attention-getting behavior, where a dog or cat has learned to feign a sickness to get attention, and fear-induced aggression, where an animal has learned to act aggressive to repel a stimulus that induces fear. On the treatment side, some emotional responses acquired through classical conditioning can be eliminated through gradual extinction, which is referred to as systematic desensitization. Natural fear responses, such as reactions to thunderstorms can be habituated. Many of the remedial approaches to problem behavior involve the conditioning of a new response that is incompatible with the undesirable behavior. This is referred to as counterconditioning. Sometimes a combination of approaches is used. A few examples of treating problem behavior are mentioned below to illustrate the effectiveness of learning principles, but not to serve as a definitive guide to behavior modification therapy.

Attention-Getting Behavior in Dogs Dogs love attention, and dogs who are sick or injured easily come to enjoy the extra attention and sympathy they receive. Some then continue acting as though they have not recovered long after they are well. Dogs playing the "sick pet game" in this way may assume a variety of medical problems, including coughing, diarrhea, vomiting, ear problems, and lameness (Tanzer and Lyons, 1977). Overloving a pet that is recovering from an illness may encourage the pet to act out the sickness again to get the reinforcing attention.

The reward for an animal displaying attention-getting behavior may shift over time. Initially a dog may bark at shadows and be rewarded by the owners for its "cute" behavior and enthusiastically be comforted. In subsequent days the dog might continue to bark at shadows but be less enthusiastically rewarded. If the dog continues to bark at shadows, the owners will tire of the noisy behavior, and the barking that was previously rewarded may instead elicit irritation and yelling. For the dog, this is still a form of attention that maintains the learned act, and it is better than being ignored.

Since some attention-getting behaviors resemble those of various illnesses, it may be necessary to determine whether the animal displays the behavior when the owners are gone. If the owners surreptitiously watch the dog while it believes they are gone, they can assess the occurrence of the behavior in their absence. If the behavior disappears, it is probably attention-getting rather than an organic problem (Hart, 1979). To eliminate the behavior, the dog must be completely ignored when the behavior occurs, and it will gradually be extinguished. Since the reinforcement ratio for such acts is usually low, the animals are not too resistant to extinction of the behavior.

Sibling Rivalry—Canine Style What would you do if you had a newborn baby and found your dog growling at the baby one day? Like most people, you would probably feel safest in banishing the dog from the room whenever the baby is around. But then you assume the dog is "jealous" and so you give it extra attention, but only when the baby is not around and competing for your attention. Now the dog learns that it is removed from the room whenever the baby is present but rewarded when the baby is away. This is your classic sibling rivalry: Get rid of the new baby. Although the baby had previously been a neutral stimulus, it becomes a conditioned stimulus for the various unpleasant responses associated with being removed from the center of social contact. Since removal of the baby is accompanied by a renewal of social contact and affection, disappearance of the baby becomes reinforcing. This problem is more easily prevented than cured, but careful control of attention to the dog can be effective in correcting the problem. The cure is that when the baby is present, affection is profusely poured on the dog, but no attention is given in the baby's absence.

Fear-Induced Aggression   Animals frequently resort to aggressive behavior when confronted with frightening stimuli from which they cannot escape. The usual approach is to attempt to desensitize the animal to the stimuli that evoke the fear. In desensitization the animal is presented initially with a very mild form of the fear-evoking stimulus, a form that barely, if at all, evokes an emotional response. This allows a degree of extinction to occur to the stimulus. Next, the intensity of the stimulus is increased slightly and extinction allowed to occur again. The process is continued until the stimulus is brought into the picture at full-blown intensity.

An example of using desensitization to deal with a Thoroughbred colt who kicked viciously whenever anyone touched his left leg is described by Voith (1979). The colt was gradually exposed to the fear-inducing stimulus by having someone move his hands toward the colt's leg. The colt was also rewarded with a bit of grain for not resisting the approaches. As the procedure progressed the handling was gradually moved down the leg. Sometimes the handler had to back off and touch the leg for only a second or two, and later extend the time of contact. The entire sequence of desensitization took three sessions of half an hour each.

Use of Punishment   In our customary interaction with animals we frequently use aversive stimulation. In the chapter on social behavior it was emphasized that punishment is necessary at times for us to maintain dominance over animals when we are the target of threats or aggressive behavior. Judicial and occasional use of punishment to reinforce our position of dominance seems to facilitate interactions with social animals especially dogs and horses. This type of punishment can be referred to as interactive punishment.

In attempting to deal with common misbehavior such as chewing up things, interactive punishment may be undesirable, because the animal's attention is focused upon us. However, remote punishment may be quite effective. A type of remote punishment is to booby trap the problem area with mousetraps. Remote punishment allows us to pair the aversive stimulus immediately with the undesirable act and to deliver the punishment within a second or less after the undesirable behavior occurs. It also allows us to remove ourselves from the process of punishment so that we do not acquire the aversive properties of a punishing stimulus.

References

ADLER, L., AND H. ADLER. 1977. Ontogeny of observational learning in the dog (*Canis familiaris*). *Develop. Psychobiol.* 10:267–272.

BALDWIN, B., AND D. INGRAM. 1967. Behavioral thermoregulation in pigs. *Physiol. Behav.* 2:15–21.

BALDWIN, B. A., AND G. B. MEESE. 1977. Sensory reinforcement and illumination preference in the domesticated pig. *Anim. Behav.* 25:497–507.

BITTERMAN, M. E. 1965. Phyletic differences in learning. *Amer. Psychol.* 20:396–410.

BITTERMAN, M. E. 1975. The comparative analysis of learning. *Science* 188:699–709.

BOLLES, R. C. 1973. The comparative psychology of learning: The selective association principle and some problems with "general" laws of learning. In *Perspectives on animal behavior*, pp. 280–306, ed. G. Bermant. Glenview, Ill.: Scott, Foresman.

BRELAND, K., AND M. BRELAND. 1961. The misbehavior of organisms. *Amer. Psychol.* 16:681–684.

BRELAND, K., AND M. BRELAND. 1966. *Animal behavior.* New York: Macmillan.

CROOK, J. H., AND J. D. GOSS-CUSTARD. 1972. Social ethology. *Ann. Rev. Psychol.* 23:277–312.

FISKE, J. C., AND G. D. POTTER. 1979. Discrimination reversal learning in yearling horses. *J. Anim. Sci.* 49:583–588.

GALEF, B. G. 1975. The social transmission of acquired behavior. *Biol. Psychiatry* 10:155–160.

GARCIA, J., W. G. HANKINS, AND K. W. RUSINIAK. 1974. Behavioral regulation of the milieu interne in man and rat. *Science* 185:824–831.

HART, B. L. 1975. Learning abilities in cats. *Feline Pract.* 5(5):10–12.

HART, B. L. 1979. Attention-getting behavior. *Canine Pract.* 6(3):10–14.

HART, B. L., AND L. A. HART. 1985. *Canine and feline behavioral therapy.* Philadelphia: Lea & Febiger.

HERBERT, J. M., AND C. M. HARSH. 1944. Observational learning by cats. The problem of imitation. *J. Comp. Psychol.* 2:81–95.

KILEY-WORTHINGTON, M., AND P. SAVAGE. 1978. Learning in dairy cattle using a device for economical management of behavior. *Appl. Anim. Ethol.* 4:119–124.

KRATZER, D., D. NETHERLAN, R. E. PULSE, AND J. P. BAKER. 1977. Maze learning in Quarter Horses. *J. Anim. Sci.* 45:896–902.

MADER, D. R., AND E. O. PRICE. 1980. Discrimination learning in horses: Effects of breed, age and social dominance. *J. Anim. Sci.* 50:962–965.

MOORE, C. L., W. G. WHITTLESTONE, M. MULLORD, P. N. PRIEST, R. KILGOUR, AND J. L. ALBRIGHT. 1975. Behavior responses of dairy cows trained to activate a feeding device. *J. Dairy Sci.* 58:1531–1535.

OLEY, N. N., AND B. M. SLOTNICK. 1970. Nesting material as a reinforcement for operant behavior in the rat. *Psychon. Sci.* 21:41–43.

OTTENBERG, P., M. STEIN, J. LEWIS, AND C. HAMILTON. 1958. Learned asthma in the guinea pig. *Psychosom. Med.* 10:395–400.

ROSENTHAL, R. 1965. *Clever Hans (the horse of Mr. von Osten).* New York: Holt.

SCOTT, J. P., AND J. L. FULLER. 1965. *Genetics and the social behavior of the dog.* Chicago: University of Chicago.

SELIGMAN, M. 1970. On the generality of the laws of learning. *Psychol. Rev.* 77:406–418.

SKINNER, B. F. 1938. *The behavior of organisms.* New York: Appleton-Century-Crofts.

SOLTYSIK, S., AND B. A. BALDWIN. 1972. The performance of goats in triple choice delayed response tasks. *Acta Neurobiol. Exp.* 32:73–86.

TANZER, H., AND N. LYONS. 1977. *Your pet isn't sick (he just wants you to think so)*. New York: Dutton.

THORNDIKE, E. L. 1964. *Animal intelligence*. New York: Hafner (originally published New York: Macmillan, 1911).

VOITH, V. L. 1979. Treatment of fear-induced aggression in a horse. *Mod. Vet. Pract.* 60:835–837.

WARREN, J. M. 1965. Primate learning in comparative perspective. In *Behavior of nonhuman primates*, vol. 1, eds. A.M. Schrier, H. F. Harlow, and F. Stollnitz. New York: Academic.

WOODS, S. C., AND P. J. KULKOSKY. 1976. Classically conditioned changes of blood glucose level. *Psychosom. Med.* 38:210–220.

ZENTALL, T. R., AND J. M. LEVINE. 1972. Observational learning and social facilitation in the rat. *Science* 178:1220–1221.

CHAPTER NINE ■ HORMONAL INFLUENCES ON BEHAVIOR

Throughout most of the history of animal domestication we have had to alter and shape the behavior of animals derived from wild stock so that their behavior is manageable under confinement and amenable to close association with humans. Much of the behavioral alteration has been done by selective breeding to reduce fear of people and enhance the traits of docility and tolerance for restraint and crowding. We take advantage of early experience and also use a variety of conditioning procedures to shape our animals' behavioral patterns to our liking. By changing an animal's hormones, especially those produced by the gonads, we can alter still further some aspects of behavior. The gonads of male animals are easily accessible to surgery, and males quickly recover from castration with little risk of infection. Thus the surgical or physiological approach to behavioral controls has been a part of domestication almost as long as the process of selective breeding.

Since hormones have profound effects on behavior, this is an area of obvious importance in understanding domestic animal behavior. Nowadays nonbreeding males of all domestic species except dogs are routinely castrated, primarily for behavioral reasons. Females intended as pets are frequently ovariectomized for behavioral reasons as well as for birth control.

The consideration of hormones and behavior involves not only the way hormones influence behavior but also the way behavioral experiences influence hormonal secretions. Our grouping, confinement, and transportation of animals affects hormones of the sex organs and adrenal glands. The concept of psychosomatic illness rests on the principle that certain experiences affect hormone secretions, which in turn affect bodily processes.

The hormones about which we have the most information relative to animal behavior are those of the pituitary gland, adrenals, and gonads, and so the discussion in this chapter focuses on these organs. It is evident, however, that hormones from other endocrine organs, such as the thyroid gland and the insulin-producing cells of the pancreas, have minor behavioral influences themselves and are influenced by behavioral events (Mason, 1968a, 1968b).

The first section of this chapter focuses on the effects that behavioral events have over the secretion of hormones. We will be concerned mostly with the pituitary and adrenal glands. The second section deals with the effects that the androgens secreted from the male gonads, especially testosterone, have on behavior, and the third section deals with the role of the ovarian hormones, estrogen and progesterone, in influencing behavior. A fourth section covers applied and practical considerations related to management practices and hormones and behavior.

■

BEHAVIORAL INFLUENCES ON HORMONE SECRETION

A number of behavioral processes and environmental factors influence the secretion of pituitary hormones. These hormones, especially the adrenocorticotropic and gonadotropic hormones, influence the secretion of hormones of endocrine systems, which in turn have major effects on physiology and behavior. A type of positive feedback may be involved in which the occurrence of certain behavioral events such as sexual activity enhances the secretion of gonadotropins, which increase the secretion of sex hormones, which in turn enhance sexual activity.

The pituitary gland is connected to the hypothalamus through the infundibulum. Secretion of hormones such as oxytocin from the posterior lobe of the pituitary is influenced by neural connections from the hypothalamus, whereas secretion of the hormones of the anterior lobe, including adrenocorticotropic hormone (ACTH), growth hormone, and the gonadotropins, is controlled by releasing factors carried from the hypothalamus to the pituitary by a vascular system. Because of the multitude of neural connections between the hypothalamus and other parts of the brain, there is a direct pathway for behavioral events to influence hormonal secretions.

Hormones of the Adrenal Gland

The adrenal gland produces two types of hormones, epinephrine from the medulla and corticosteroids from the cortex. Behavioral events that influence the secretion of hormones of the adrenal medulla do so through neural connections innervating the adrenal medulla.

The secretion of epinephrine has general activational effects and has long been known to respond to psychological influences. The best-understood influence is the marked elevation in the secretion of epinephrine that results from novel or unpredictable situations. Since this hormone results in an arousal of the central nervous system and a general physiological activation for muscular exertion, this type of physiological activation is usually advantageous in times of emergency.

Under domestic conditions we tend to subject animals to novel environments and unknown situations rather frequently. This may lead to excessive activation of the adrenal medulla and predispose some animals to the adverse effects of excessive and continuous activation.

The same type of unpredictable, threatening, or novel situations that activate the adrenal medulla may also lead to stimulation of the adrenal cortex. However, the control of cortical hormones is more complicated because of existing homeostatic mechanisms. Like other hormone systems, secretion of these corticosteroids is controlled by a negative feedback mechanism whereby the corticosteroids act on the hypothalamus to suppress the secretion of the ACTH releasing factor. If corticosteroid secretion decreases, ACTH levels increase, and if steroid levels rise, ACTH levels fall. Under conditions of physical or emotional stress, the negative feedback mechanism is overridden, and corticosteroid levels increase rapidly to help prepare the animal for the stressful situation by a variety of biochemical processes.

The corticosteroids can be classified as to principal biological action. The mineralocorticoids, of which aldosterone is the most important, function to maintain optimum sodium concentration in the blood and extracellular fluid. The glucocorticoids, of which cortisol and corticosterone are the most important, play a vital role in carbohydrate metabolism. So far, the primary link of behavior involves the glucocorticoids rather than the mineralocorticoids.

The adrenal cortex of humans and most domestic animals secretes both cortisol and corticosterone, with cortisol usually having the highest secretory rate. Most of the behavioral work has been done in rats, and in this species almost all of the glucocorticoid secretion is of corticosterone. The glucocorticoids promote the formation of glucose from noncarbohydrate precursors. They maintain extracellular fluid volume by preventing the shift of water into cells (helpful in case of hemorrhage from fight wounds). The glucocorticoids have strong anti-inflammatory effects that seem to diminish allergic responses. Further, when the animal is confronted with an opponent in a fight, the anti-inflammatory effects of the glucocorticoid secretion would temporarily attenuate the pain, enabling the animal to continue to fight. When an organism is not subjected to stress, the glucocorticoids are maintained at a level necessary for normal bodily processes. Lack of glucocorticoids resulting from adrenal insufficiency can lead to muscle weakness, hypertension, and reduced ability to withstand stress.

The increased secretion of glucocorticoids following physical stress, which was labeled the general adaptation syndrome by Selye (1950), helps the animal to survive physical stress. Under ideal conditions, when the stress is relatively brief, glucocorticoid secretion subsides rapidly. If the corticoid secretion is maintained for a long period at high levels, as in prolonged physical stress, the effects can be very detrimental. There may be excessive breakdown of body protein, excessive tissue edema, and interruption of growth. The animal may be rendered more susceptible to some disease processes. One reliable measure of excessive corticoid output is hypertrophy of the adrenal cortex resulting from excessive stimulation of the cortex.

Both ACTH and the corticosteroids seem to have some direct effects on behavior. Evidence indicates that ACTH enhances generalized excitability or anxiety whereas the corticosteroids act to reduce excitability to a more normal level (De Wied, 1974; Weiss et al., 1970).

Behavior and Corticosteroid Secretion   One of the phenomena that help animals in the wild to survive physical stress, such as fights with predators or conspecifics, is that behavioral experiences associated with anticipation of the stress—seeing the opponent or predator, for example—evoke a burst of secretory activity from the adrenal cortex. Thus an animal's physiological mechanisms for dealing with stress are operating before the physical demands.

On the domestic scene we seldom allow fights or other types of physical trauma to occur and endanger animal lives. Yet the mechanism for behavioral activation of the system still operates. In addition to the presence of predators or adversaries, we know that situations beset with unpredictability or uncertainty also activate the adrenal cortex. In the domestic environment we have greatly reduced physical stress and have introduced a different type of novelty and unpredictability than occurs in the natural environment.

Often the first exposure to a new environment or test situation evokes the highest ACTH response. Situations fraught with fear, or in which long established rules are suddenly changed, or in which behavior that was previously effective no longer accomplishes the task may lead to unusually high levels of ACTH production. Just as prolonged physical stress is detrimental, psychological conditions resulting in constant fear or anxiety may lead to general debilitation, weakness, gastric ulceration, and cardiovascular disturbances because of the prolonged and excessive output of ACTH and glucocorticoids.

Not surprisingly, several studies have shown that excessive crowding can lead to impaired health and reproductive failure. It is generally assumed that crowding causes an excessive number of social encounters and competitive interactions, resulting in excessive fear and anxiety. In crowded animals, it has been proposed that with excessive ACTH production, the overactive adrenal cortex produces sufficient steroids with androgenic properties to suppress gonadotropin production by the pituitary, resulting in impairment of reproductive function.

Infantile Stimulation and ACTH Production   In Chapter 7 it was pointed out that stress on the neonate in the form of handling for just a few minutes a day may have influences on the growing and adult animal. Neonatal stress results in more rapid growth of various organ systems. The adult that has been stressed in infancy usually weighs more and has a slightly larger brain than nonstressed animals. There are indications that the stressed animals are more resistant to some disease processes. One of the differences between stimulated and nonstimulated animals, when tested in adulthood, is excessive reactivity of the nonstimulated animals to novel, but not physically threatening, situations. The nonstimulated rat freezes, urinates, and defecates in a novel environment whereas the stimulated rat freely and unemotionally explores the new situation.

The behavioral and physiological effects of neonatal stress point to the involvement of ACTH and adrenal cortical hormones. A mechanism by which the pattern of ACTH secretion is established or influenced by early stimulation has been proposed by Levine (1969). It has been shown that stimulation of newborn rats can activate the ACTH-steroidogenesis system during the first few days of life. Levine postulates that the brain of the rat is particularly sensitive to corticosterone, and that this hormone permanently modifies the neural development of structures determining the pattern of release of ACTH later in the organism's life. According to this theory, infantile stimulation promotes a type of neural development that later allows the release of ACTH to be graded so that the amount of corticosteroid secreted matches the degree of actual physical stress to which the organism is subjected. In the nonhandled animal, in which ACTH or corticosteroid secretion is not evoked during the critical period, optimal development of the nervous system, allowing a graded release of ACTH, would not occur.

Behavioral Influences on Oxytocin Secretion

The best-known effects of oxytocin are to induce milk letdown in lactating females and to stimulate uterine contractions during copulation and parturition. In lactating cows oxytocin is released in transient bursts before and during milking. This hormone facilitates milk letdown action by causing a contraction of the myoepithelial cells of the mammary gland alveoli. Oxytocin release by the posterior pituitary is part of a neuroendocrine reflex that is normally activated by tactile stimulation of the mammary gland through sucking or milking.

In dairy cattle this milk letdown reflex is susceptible to conditioning such that the sound or sight of a farmer moving milking cans or in other ways preparing for milking frequently causes the milk to be let down before the udder is even touched. Visual and auditory stimuli that are closely associated with milking or nursing and that cause milk letdown may result in the buildup of pressure in the mammary gland to such an extent that milk will drain through the teat orifice.

Just as some stimuli may induce milk ejection through conditioning, other stimuli, especially those associated with physical or emotional trauma, may reduce milk ejection. This has been illustrated in rats in an experiment showing that a variety of stimuli, namely the odor of oil of peppermint, the flashing of bright lights, and human conversation, reduced the milk yield for lactating females (Grosvenor and Mena, 1967). The inhibitory effects of stimuli on milk yield are believed to be due to the release of epinephrine and norepinephrine from the adrenal medulla.

The milk ejection reflex in women has played a notable role in the history of art and literature. The Greek legend about the origin of the Milky Way, recounted by Folley (1969), reminds us that the ancients had an interest in this subject. The Greek god Zeus, wishing to immortalize Hercules, his son by a mortal woman, presented the infant to the breast of the sleeping goddess Hera. The infant, after becoming satiated, stopped suckling, but milk continued to flow from Hera's breasts upward into the heavens, forming the galaxy we call the Milky Way.

There is some evidence that oxytocin is released during copulation in both males and females. In the female oxytocin can act to increase contractions of the uterus, which is believed to facilitate sperm transport. In the male oxytocin can affect smooth muscle along the male genital system and help to replenish sperm reserves, which would increase sperm counts on subsequent ejaculations (Sharma and Hays, 1968; Voglmayr, 1975).

Behavioral Influences on the Gonadal Hormones

Hormones secreted by the pituitary that influence the gonads are FSH (follicle-stimulating hormone), LH (luteinizing hormone), and prolactin. These three gonadotropic hormones control the morphological and physiological aspects of ovarian function in the female and hence are related to changes in estrogen and progesterone production. In the male, FSH and LH are more or less tonically secreted, compared with the cyclic secretion of these hormones in the female. Testosterone production by the interstitial cells of the testes is stimulated by LH, and FSH stimulates the production of sperm cells.

The hormones FSH, LH, and prolactin apparently do not have major direct influences on behavior, but produce most of their behavioral effects indirectly by causing changes in gonadal hormone production. The gonadal hormones, of course, have profound effects on the functioning of the reproductive system and on sexual behavior.

Females of species that are seasonal in reproductive function respond primarily to day length. This effect is mediated through connections of the visual system to the hypothalamus, with effects relayed to the gonads through the hypothalamic-pituitary connections.

Effect of Males on Females   A stimulus that affects the onset of sexual activity in females that are seasonal breeders is the presence of the male. For example, female sheep come into estrus earlier in the breeding season if rams are included in the flocks (Fletcher and Lindsay, 1971; Lindsay et al., 1975). Evidence shows that this effect on the female is due to olfactory stimulation from the smell of rams (Morgan et al., 1972).

Another effect of the presence of males is acceleration of the onset of puberty. The timing of the onset of puberty is basically a hypothalamic function, and stimuli from males can affect this as they can the seasonal onset of sexual activity. The first demonstration of such an effect was in mice, in which it was found that odors produced by males accelerated female sexual maturation (Vandenbergh, 1969*a*). Since then it has been found that the introduction of a boar to prepubertal gilts results in estrous activity sooner than without the boar present. There is also evidence that the onset of ovarian activity in heifers is slightly accelerated by the presence of sexually mature bulls (Izard and Vandenbergh, 1982).

One other behavioral effect on reproductive function seen in females relates to the timing of ovulation. Ovulation is induced by a surge in the production of LH. In most domestic species this occurs as a function of intrinsic hypothalamic timing. In a few species the surge of LH occurs as a result of copulation. This is true in cats, rabbits, ferrets, and mink. These animals are referred to as reflex ovulators because copulation activates release of LH through a neuroendocrine reflex. There is evidence that even in spontaneous ovulators, however, copulation or just the presence of a male will hasten the onset of ovulation (Jochle, 1975).

Effects of Females on Males   If males can have an effect on the reproductive functioning of females, it is logical that the presence of females may have physiological effects on males. Effects of females on male seasonal changes is rather limited but it has been shown in monkeys. Male rhesus monkeys on island colonies near Puerto Rico, which breed on a seasonal basis, were made sexually active during the nonbreeding season by the exposure to females artificially brought into estrus (Vandenbergh 1969*b*). The same effect has been shown in hamsters (Vandenbergh, 1977).

Females also sometimes elicit a surge in testosterone production in males just before, during, or after copulation. This has been documented in bulls, rams and boars, as well as in laboratory animals (Katongole et al., 1971; Liptrap and Raeside, 1978; Sanford et al., 1977). Illustrations of this effect are shown in Figure 9-1. Simply exposing a male to a female and creating sexual excitement has the same effect. The surge of testosterone is usually preceded by a surge in LH production. The hormone-inducing effect of the presence of females is most readily seen in older, more experienced males.

Whether these transient bursts in testosterone in farm animals have a physiological or behavioral role in reproduction is not clear. There is evidence of a significant cumulative effect of repeated exposures to females. In rats the spontaneous atrophy of the reproductive system that occurs with age is attenuated by housing males with females and allowing copulation ad libitum (Drori and Folman, 1964; Thomas and Neiman, 1968).

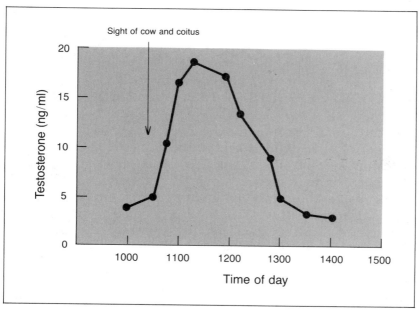

Figure 9-1 Example of the effects of the sight of a cow and coitus on blood concentration of testosterone in a bull: The peak in testosterone is usually preceded by a peak in luteinizing hormone (LH). If blood testosterone concentration is already high, the sight of a cow or coitus does not cause a further rise in testosterone. (Modified from Katongole et al., 1971.)

One of the more interesting effects of behavioral experiences on testosterone secretion is that following social stress. In man it has been observed that stressful events such as going through officer candidate school or basic combat training causes a fall in testosterone secretion. The period after winning a tennis match or getting a medical school diploma is likely to be characterized by an increase in testosterone in human males (Mazur and Lamb, 1980). These phenomena are applicable to animals as well. When male rhesus monkeys are defeated in a social encounter, they experience a fall in testosterone. If they are lucky enough to become dominant, they are likely to undergo a marked increase in testosterone secretion (Rose et al., 1975). To the extent that testosterone alters a monkey's aggressive behavior or changes its stimulus value as an animal to attack or avoid, these effects are adaptive. An increase in testosterone for the winner helps him to maintain his dominant position, and the fall in testosterone decreases the likelihood that the loser will engage in a fight in the near future. How widespread these phenomena are outside of monkeys is not known.

■

BEHAVIORAL EFFECTS OF TESTICULAR HORMONES

One of the most frequently employed means of changing the behavior of male animals kept in the domestic environment is by castration. Since the testicles are the primary source of male hormones, or androgens, removing them reduces the available androgens to the relatively small quantity produced by the adrenal glands. Castration may be performed prepubertally or postpubertally. Although castration is quite effective in altering certain specific aspects of behavior, elimination of gonadal hormones does not always produce the results we expect. In this section we shall examine the basis for determining which behavioral patterns will be altered and some of the species differences in reaction to castration.

The Concept of Sexual Dimorphism

Males and females of all domestic species except the dog differ markedly in size. Males are larger and are characterized by enhancement of some portions of the musculoskeletal system. Neck and shoulder muscles are usually larger. In sheep males have horns but females do not. In cattle the male's horns are larger. These sexually dimorphic differences have been related to male competition for females and the selective advantage of appearing strong and potentially dangerous to a conspecific male opponent. Wolves, which are the ancestors of dogs, are monogamous and live a life style characterized by reduced competition for females. Correspondingly, there is little morphological sexual dimorphism between the sexes.

In the previous chapters on feeding, social, sexual, and maternal behavior, we examined some behavioral differences between males and females. Males eat more than females. With few exceptions, males are more aggressive. This is true of interactions between adults, but it may also be true of other types of aggressive behavior. Aggressive behavior of all kinds is more frequently a clinical problem in male dogs than female dogs (Voith, 1979). Urine marking is commoner in male dogs and cats than in females. Flehmen in ungulates is performed much more frequently in males than females. Of course, patterns of sexual and parental behavior differ between males and females. Males of the common livestock species do not show any degree of parental behavior.

The neural circuitry underlying the behavioral patterns that are typical of males or females appear to exist in both sexes. Females, for example, can fight as viciously as males, and females can also display flehmen, urine marking, and male patterns of sexual behavior. The circuitry underlying the behavior of the opposite sex appears to be held under neuroinhibitory influences, but the behavior typical of the opposite sex may be shown occasionally. This is best documented with regard to sexual behavior (Hart and Leedy, 1983). Therefore, behavioral differences between males and females are of tendency or degree; they are not absolute.

With some behavioral patterns the differences between the sexes are reduced by removal of the testes in adult males. Sexual behavior, as well as intermale fighting, is generally reduced by castration. For the most part, the administration of testosterone to ovariectomized females and of estrogen to castrated males does not change sexually dimorphic behavioral characteristics accordingly and, in fact, testosterone may activate female behavior in females and estrogen may activate male behavior in castrated males.

Behavioral differences between males and females are not limited to just those patterns that can be observed in adults. It is well documented that play in juveniles, when there is little secretion of gonadal hormones in either sex, is sexually dimorphic. Male puppies are more aggressive and mount each other more frequently than females (Bekoff, 1974). The urination posture of male and female puppies differs as well (Figure 9-2). A study of juvenile play in sheep reveals that male lambs are much more likely to mount other lambs than are females and are also more likely to butt each other than are females. On the other hand, there is a slight predominance of gamboling (stiff-legged jumping motions) in females (Sachs and Harris, 1978).

Androgen Secretion and Male Behavior

Although it is believed that the chief male sex hormone that influences behavior is testosterone, another androgen, andros-

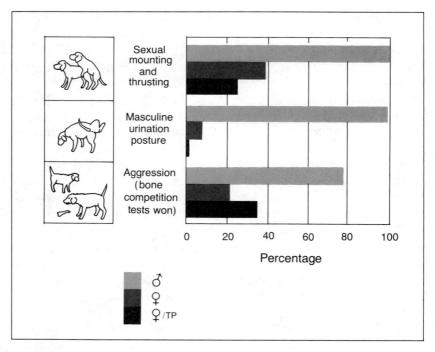

Figure 9-2 Sexually dimorphic behavior: Sexual mounting and thrusting, urination posture, and aggression are examples of behavioral patterns that are sexually dimorphic in dogs. Shown here are graphs representing the performance of the masculine forms of the behavior, illustrating that differences are of frequency or degree and not absolute. Administration of testosterone propionate to females (♀/TP) has little or no effect on their behavior. The dependent variables for each graph differ. For mounting, the percentage of adult males and females in each experimental group displaying mounting and thrusting is depicted. For urination posture, the percentage of all urination acts in which the leg-lift posture was displayed is shown. For aggression, the percentage of bone competition tests won by males or females is shown. (Data from Beach, 1974; Beach et al., 1982; Beach and Kuehn, 1970).

tenedione, is involved to a lesser extent. Both androgens are produced by the interstitial cells (of Leydig) in the testes. When males have been castrated, the commonest replacement form of androgen used both clinically and experimentally to restore male behavior is some form of testosterone.

The production of testosterone is governed to some degree by negative feedback effects of testosterone on the hypothalamus. Presumably a decrease in androgen production signals the hypothalamus to secrete more releasing factor, which increases

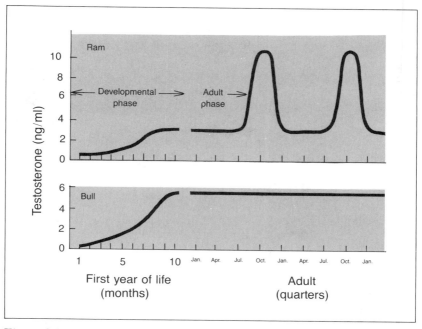

Figure 9-3 Developmental patterns of testosterone secretion in rams and bulls as measured by the concentration of testosterone in blood plasma: After adulthood is reached rams undergo marked seasonal fluctuations in the hormone secretion, whereas in bulls testosterone secretion is the same year-round. Both rams and bulls have daily fluctuations in testosterone levels that are not represented in the graph. (Based on data from Lee et al., 1976; Secchiari et al., 1976; Sanford et al., 1977.)

the output of the luteinizing hormone from the pituitary, resulting in more testosterone production. A rise in the blood concentration of testosterone affects hypothalamic neurons in the opposite way and, through a reduction in the releasing factor, luteinizing hormone secretion is reduced, resulting in a decrease in testosterone production.

The developmental pattern of testosterone secretion in males, as represented by the concentration of the hormone in blood plasma, is shown schematically for the ram and bull in Figure 9-3. These graphs illustrate the increase in testosterone that occurs at puberty in the males of all mammalian species.

In adult males of virtually all species studied there are marked daily fluctuations in testosterone levels of the type illustrated in Figure 9-4 for the bull. In most species studied the fluctuations can be as great as tenfold. Usually the peaks in tes-

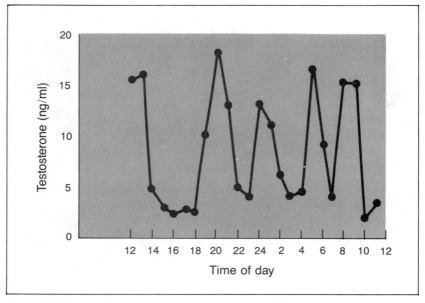

Figure 9-4 Daily fluctuations in testosterone blood concentration of bulls throughout the day: Each peak in testosterone is usually preceded by a peak in luteinizing hormone. (Modified from Katongole et al., 1971.)

tosterone concentration are preceded by bursts in production of luteinizing hormone. As mentioned above, exposure of a male to a female often produces a surge of testosterone production. Like the random fluctuations in secretions, the surge in testosterone produced by sexual excitement is usually preceded by luteinizing hormone release.

The fact that there are fluctuations raises the question of what is the minimal level of testosterone needed to maintain behavior. In rams that had been castrated before puberty (wethers) graded doses of testosterone were used to test the threshold required for mating activity. The dose that stimulated the complete mating response, including intromission and ejaculation, resulted in relatively low plasma testosterone levels, comparable to those seen in rams during the nonbreeding season. A much lower dose was sufficient to elicit mounting behavior without intromission or ejaculation (D'Occhio and Brooks, 1982). Similarly, work with rats suggests that average testosterone levels are three to four times higher than the levels needed to maintain sexual behavior (Damassa et al., 1977). These studies may ex-

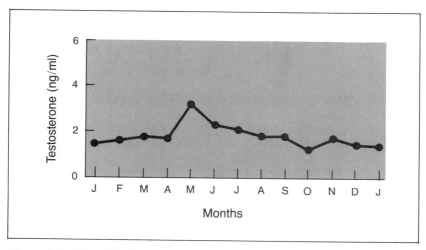

Figure 9-5 Monthly changes in blood plasma concentration of testosterone in stallions show only moderate seasonal effects. (From data presented by Berndtson et al., 1974.)

plain why there is no simple relation between plasma testosterone level and mating activity in male mammals.

Most mammalian species are seasonal breeders in that females come into estrus in only a certain part of the year. The adaptive value of such a process is that the young are born at the most favorable time. It has been observed in several wild species that there are also seasonal changes in testosterone in males that correspond to the seasonal cycle of females. In many species of deer, for example, intermale aggressive behavior is markedly intensified during the rut, or breeding season, but the animals are relatively nonaggressive during other times of the year. These behavioral changes are attributed to seasonal changes in testosterone.

The domestic species with the most pronounced seasonal fluctuations in testosterone secretion are sheep and goats. Figure 9-3 illustrates the general pattern and time scale of seasonal changes in the ram compared with those of the bull. The ram's off-season baseline level of testosterone is lower than that of bulls, but its levels exceed those of the bull during the breeding season. Monthly changes in testosterone concentration in the stallion are more moderate than those of the ram (Figure 9-5). Through artifical breeding and selection, the seasonality of androgen production has been attenuated or eliminated in some species, as with cattle (Figure 9-3).

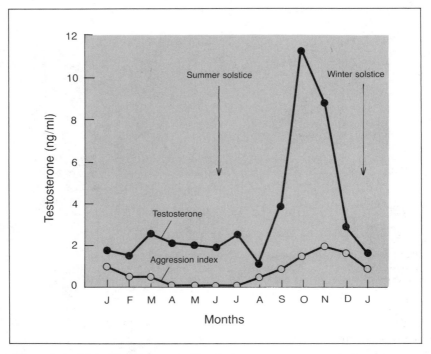

Figure 9-6 Monthly changes in blood plasma concentration of testosterone and aggression in adult Soay rams show seasonal effects. The aggression index ranged from 0 to 4. (Modified from Lincoln and Davidson, 1977.)

One of the ancient breeds of sheep, the Soay sheep of Scotland, has been studied because of the marked seasonality in their mating activity. In Figure 9-6 we see a change in testosterone production as a function of season. The relation of aggressive behavior to this change in testosterone is also evident on the graph. As a measure of aggression the investigators used a mechanical counter that recorded each time the ram butted against his pen railing in an attempt to fight with the ram in the neighboring pen.

The same investigators determined that testosterone production could be influenced artificially by maintaining the Soay sheep under artificially long days and suddenly switching them to short days. The results schematically portrayed in Figure 9-7 illustrate the time lag from the onset of short days to the onset of the testosterone surge and the subsequent increase in fre-

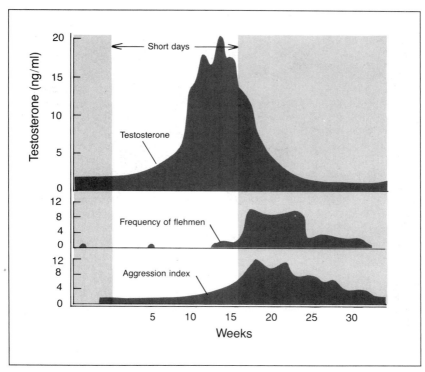

Figure 9-7   Measures of blood plasma testosterone concentration, weekly counts of frequency of flehmen, and aggression index in Soay rams housed under artificial lighting and switched from long days to short days. (Modified from Lincoln and Davidson, 1977.)

quency of sexual behavior, aggression and flehmen (Lincoln and Davidson, 1977).

In cats, stallions, and rams of the modern breeds of sheep, copulation will occur during the periods of low testosterone production but the level of sexual activity is diminished (Berndtson et al., 1974; Sanford et al., 1977).

Individual Differences in Androgen Concentration   Some investigators have been concerned with determining if the natural differences in sexual activity one finds in animals are a reflection of differences in androgen blood levels. Generally it has been found that by giving supranormal doses of androgen, the degree of sexual activity in laboratory rodents can be increased to some extent but that individual differences cannot be overcome (Beach and Fowler, 1959; Damassa et al., 1977).

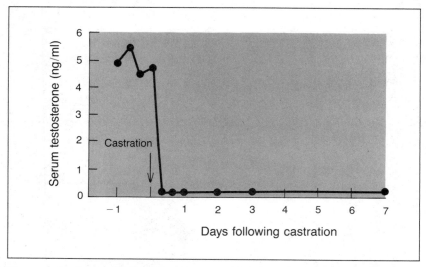

Figure 9-8 Example of the time course of reduction of testosterone in blood serum following castration (in this instance taken from a study in cats): The three points immediately preceding and following castration represent mean levels in blood samples taken at eight-hour intervals. (From Hart, 1979.)

Several experiments have shown that there is no relation between sexual activity and testosterone levels. The usual procedure is to study animals ranging from low to high in sexual activity. All the animals are then castrated and tested until they stop mating. They are then given the same amount of testosterone on a daily basis until sexual activity returns. In subsequent tests for sexual behavior it has been found that males return to their respective preoperative levels of sexual activity (Grunt and Young, 1952; Larsson, 1966; Harding and Feder, 1976).

When adult males are castrated, all the behavioral patterns that are androgen-dependent continue to be displayed for a period of time after androgens have disappeared from the blood. The persistence of behavior for weeks, months, or sometimes years is a reflection of the capability of the central nervous system to continue to mediate this behavior without androgen support. Following castration, testosterone almost completely disappears from the bloodstream within hours in every species examined (Figure 9-8). The persistence is not due to adrenal gland secretion or to other sources of androgen. It has been

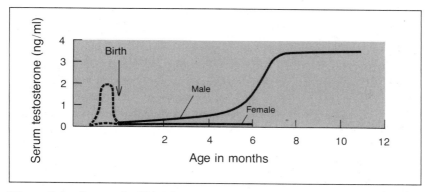

Figure 9-9 Portrayal of blood serum testosterone concentration in dogs from late prenatal life to adulthood: The concentrations represented from birth to adulthood are based on actual measurements (Hart and Ladewig, 1979, 1980). Prenatal concentrations for males and females are hypothetical and based on data from monkeys and other species (Hart, 1979).

shown in hamsters, rats, and dogs that sexual behavior in castrated males persists even after the animals are adrenalectomized (Beach, 1970; Bloch and Davidson, 1968; Warren and Aronson, 1956).

Ontogenetic Changes The testicle does not produce androgen at the same rate throughout an animal's life. Before or just after birth in males there is a surge of production of testosterone that is responsible for the differentiation of the central nervous system of the neonate toward a masculine pattern. After this neonatal period the testes produce very little androgen until the time of puberty, when androgen is again produced in relatively large amounts. The developmental pattern of testosterone secretion in all domestic animals has not been worked out yet, but it is reasonable to expect the general picture of a prenatal surge in secretion followed by a period of low secretion, followed in turn by a pubertal increase in secretion. An example of how such a developmental pattern of testosterone secretion presumably looks for the dog is shown in Figure 9-9.

At puberty increased androgen production in males stimulates the development of a number of morphological characteristics such as a powerful musculoskeletal system, growth of the penis, and maturation of accessory sexual organs. This pubertal surge in androgen production is also responsible for the onset of some masculine patterns of behavior such as sexual and aggressive behavior.

Action of Androgen on Target Tissues   One way in which hormones influence behavior is by their effect on the central nervous system. This effect probably involves the direct action of hormones on some brain cells. Crystalline testosterone implanted into neural tissue of the anterior hypothalamus and preoptic region will stimulate sexual behavior in sexually inactive castrated male rats (Davidson, 1966; Smith et al., 1977). With a special technique referred to as autoradiography, the uptake of radioactive testosterone has been demonstrated in individual neurons (Pfaff, 1968; Stumpf, 1971). Androgens are also concentrated by spinal neurons, and there is evidence that testosterone affects spinal neurons involved in the mediation of sexual reflexes (Breedlove and Arnold, 1980; Hart, 1978).

There is considerable evidence that testosterone has its primary effects on brain cells through metabolism (aromatization) to estrogen. Another metabolite of testosterone, dihydrotestosterone, appears to play a minor role in activating brain cells but is important in stimulating the spinal neurons involved in sexual reflexes and in stimulating the growth of the accessory sex organs. Typical of experiments supporting the concept that the estrogenic metabolite of testosterone is responsible for the effects of testosterone in the brain are studies revealing that estrogen, but not dihydrotestosterone, can activate mounting, copulation, and even aggressive behavior in castrated male sheep and deer (Fletcher and Short, 1974; Parrott, 1978).

To complete the picture of hormonal influences on the development of sexual behavior, the female should be mentioned at this point. It is currently believed that the prenatal or neonatal ovary does not secrete significant amounts of estrogen or progesterone. Therefore, the development of feminine behavioral patterns cannot be attributed to the presence of ovarian sex hormones. Females ovariectomized on the day of birth, when tested as adults after being injected with estrogen, show behavior identical to that of normal females in estrus.

Just as males are normally masculinized by testosterone secretion prenatally or neonatally, so females can be masculinized experimentally by injections of testosterone at this time. In sheep, for example, behavioral masculinization of female fetuses can occur if pregnant ewes are given injections of testosterone about 50 to 100 days after conception. Such masculinized females will mount normal females in estrus, display malelike courtship behavior, and perform flehmen after investigating another female (Clarke, 1977).

Female dogs that are subjected to androgen in fetal life through injections of the pregnant female also display various aspects of malelike sexual behavior when tested with normal females in adulthood (Beach et al., 1972). The critical period in fetal life for the masculinization of female dogs is closer to parturition than in sheep, which is what one would expect given that dogs are born in a more immature state than sheep.

The neonatal masculinization or feminization that takes place in the nervous system also affects sexually dimorphic behavioral patterns other than sexual behavior. In rats and mice, neonatal castration of males results in less fighting in adulthood, and androgenization of neonatal females makes them more aggressive in adulthood. The same effects are seen in sheep and dogs, with androgenized females being more aggressive toward other females than are normal females (Beach et al., 1982; Clarke, 1977).

In sheep and goats there are pronounced sex differences in the urination posture, with females showing the squatting posture and males standing somewhat stretched out. The squatting posture of females serves to keep the body from being contaminated by urine. This is not a concern of males, of course. Female sheep that are androgenized by prenatal injections of testosterone to the pregnant female tend not to display the female postures and urinate more like males (Clarke et al., 1976). In dogs not only does the posture differ between males and females, but males urinate much more frequently because urine is used regularly as a scent marking substance. Treatment of female dogs prenatally with testosterone not only induces them to display the male urination posture but to urinate much more frequently and in a scent marking fashion as do males (Beach, 1974).

In comparing androgenic effects on the urination postures of sheep and dogs, we see an interesting contrast. In sheep androgenization suppresses the squatting posture. The result is a more passive type of urination characteristic of males. In dogs androgenization not only suppresses the femalelike squatting posture but also induces the even more active leg-lift scent marking behavior. The specific effects make sense for each species.

Urination posture in sheep is basically related to sanitary eliminative behavior and urination posture in dogs relates not only
to the sanitary function but to social marking behavior. In both
instances there are species-specific needs for sex-specific behavior, and these are brought about by the developmental influences of testosterone.

Male and female urination postures differ in sheep and dogs
even in the juvenile period, during which the gonads of both
sexes are quiescent and producing no hormone in appreciable
amounts. This is a reflection, of course, of the previous prenatal
organizational effects of testosterone. For sanitary reasons it is
important for juvenile males and females to differ in urination
posture. The juvenile posture is not simply a step on the way to
the adult posture. After puberty sociosexual interactions become
important to male dogs, and we see the effects of postpubertal
secretion of testosterone in bringing on leg-lift urine marking
behavior.

Virtually all of the behaviors that are sexually dimorphic in
rodents, including general activity, pain sensitivity, taste sensitivity, food intake, body weight regulation, learning of certain
types of mazes, taste aversions, and learning performance on
certain schedules of reinforcement can be affected by prenatal
or early neonatal manipulation of androgen (Beatty, 1979).

In summary, it appears that all behavioral differences between males and females are a function of androgenic effects
on the central nervous system coupled with androgenic influences on the peripheral organs that are used in performing behavior, including the genitalia, scent marking glands, antlers,
horns, and so forth (Hart, 1974, 1978).

It is time to emphasize again that the concept of a masculine
or feminine nervous system—and hence behavior—should not
be taken to imply that a normal male is incapable of exhibiting
a typical feminine response or that a normal female is unable
to display characteristically male behavior. The difference is
rather one of tendency or probability. Normal females can and
do mount other females in estrus and do at times exhibit fierce
aggressive responses, but they have a lower tendency to display
these responses than males. Once a nervous system has developed along a masculine or a feminine line, which presumably
occurs during one or more critical developmental periods, it
theoretically cannot be reversed by subsequent hormonal manipulation. As we have seen, some sexually dimorphic patterns,

such as the type of juvenile play and urination postures, appear without any apparent postpubertal activational effects of sex hormones. Other sexually dimorphic patterns of the adults do not appear until after the postpubertal increase in sex hormone production.

Incidental Masculinization of Females A naturally occurring casualty of the prenatal masculinizing process is the freemartin heifer. If a female shares the uterus with a male fetus, she is likely to be subjected not only to a diffusion of androgen from the male but apparently even to transfer of testosterone-producing cells from the male's gonads. The freemartin's genitalia are masculinized and, as a young animal, the freemartin heifer displays masculine behavior. This is particularly noticeable in her tendency to mount other cattle and to be more aggressive.

Among dogs and cats one occasionally sees normal females show a pronounced tendency to display one or more male behavioral patterns. Urine spraying by female cats and leg-lift urine marking by female dogs are two examples often mentioned. From what we know about developmental aspects of sociosexual behavior, pronounced malelike behavior is very difficult to evoke in normal adult females even if they are given injections of testosterone. Therefore, the appearance of male behavior in such females suggests that the normal female may have been prenatally androgenized in some way. Some observations in rodents suggests how this may come about.

If female rat or mouse fetuses are situated in the uterus between two male fetuses, they are subject to diffusion of testosterone from the males. Such females have an enhanced tendency to show male sexual behavior as adults (Clemens, 1974; vom Saal, 1981). The work on mice shows that female fetuses situated between two male fetuses have higher serum levels of testosterone than female fetuses situated between females. The females that developed in utero between males urine-mark more than normal females as adults, and they are more aggressive toward other females. Yet their reproductive system remains normally femalelike, and they are perfectly fertile. The females that developed in utero between two female fetuses appear to be more attractive and sexually arousing to males. Investigators working with rats propose that normally there is a variation in sexual characteristics of female mice traceable, in part, to differential proximity to male fetuses in intrauterine life. This variation, and hence the prenatal androgenization effect, could be adaptive to the parents because offspring might thrive differentially under diverse ecological conditions. The more androgenized females would have the advantage when population density is high because they are more aggressive toward females, fiercely defend their young, and enter puberty sooner when living under crowded conditions. The nonandrogenized females would have the advantage when population density is low because they are more highly preferred by males and enter puberty sooner under noncrowded conditions. Thus in each litter there are likely to be female offspring matched for each environmental extreme.

With the experiments on rodents as background, we might imagine that the female cats that engage in urine spraying behavior and the female dogs that take up mounting or the leg-lift urination posture are those that were androgenized in utero by male wombmates. However, recent information on cats revealing no relationship between intrauterine proximity to males and the occurrence of urine spraying in females does not support this notion (Hart and Cooper, 1984).

Behavioral Effects of Castration

Androgens have a number of morphological, physiological, and behavioral influences in adult males. The hormones maintain anabolic processes in some major muscle groups and they have effects on metabolic rates and on liver enzymes. Behavioral influences include activation of sexual, aggressive, and scent marking behavior in some species. These sexually dimorphic patterns are the behaviors most likely to be affected by castration. There are other behavioral influences of castration that are not as well documented. Castration reduces spontaneous or general activity in rats, for example (Stern and Murphy, 1971).

If castration is performed on males after the critical neonatal developmental period, but before the postpubertal increase in androgen production, some elements of masculine behavior may never be shown. This is one of the primary reasons males of all domestic species, except dogs, are routinely castrated prepubertally. However, such animals have undergone the perinatal effects of androgen secretion and hence still have the potential for masculine behavior; for if given androgen any time later in life, they can display typical masculine behavior.

There are numerous examples of the potential of castrated males to show adult male behavior. One interesting case is a report of a gelding that was implanted subcutaneously with testosterone and used as a teaser for detecting estrus in mares. Seven days after being given the implant he displayed strong sexual interest in the mares, and the mares displayed signs of receptivity to him. About two weeks after removal of the implant the behavior subsided (Swift, 1972).

Since the pubertal increase in androgen production stimulates masculine musculoskeletal growth, prepubertal castration reduces masculine growth in the musculoskeletal system. Once the age of puberty is past, administration of testosterone does not bring on additional growth. A procedure by which one might separate the morphological from the behavioral changes that usually go together would be to allow the animal to develop both morphological and behavioral characteristics and then perform the castration. Behavior may thus be altered, but the morphological characteristics would not revert back to the prepubertal condition.

There seems to be the general impression among animal people that prepubertal castration has more profound effects on masculine behavior than postpubertal castration. However, the evidence does not support this concept. We have already noted that as long as male animals have undergone the perinatal effects of androgen secretion, they still have the potential for the display of masculine behavior at any time later in life. This is particularly evident in the two companion animal species, horses and cats, that are routinely castrated before puberty to prevent the occurrence of undesirable masculine behavior. Geldings are observed to occasionally display stallion-like sexual behavior and aggressiveness toward horses or people. Male cats that are castrated before puberty may take up the practice of urine spraying later in life. Two recently completed clinical surveys of problem behavior in prepubertally castrated horses and cats have revealed that the occurrence of objectionable male behavior in these animals as adults is not significantly less than the occurrence of the behavior in male animals castrated as adults. Figure 9-10 illustrates that the display of sexual behavior, aggression toward horses, and aggression toward people in prepubertally castrated geldings is about the same as the occurrence of these behavioral patterns in stallions that are castrated in adulthood (Line et al., 1984). In a survey on male cats, it was found that the percentage

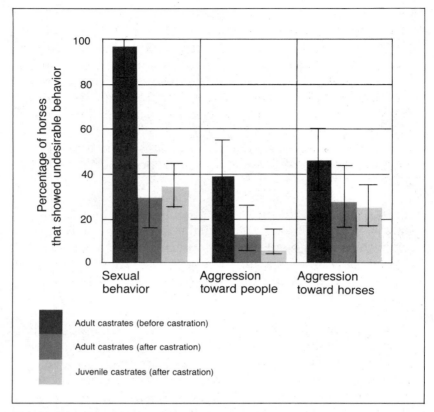

Figure 9-10 Comparison of the behavioral effects of prepubertal and postpubertal castration of male horses on problem aggression directed toward people or other horses and on sexual behavior directed toward mares: There was no significant difference between adult and juvenile castrates in any of the behavioral patterns. (From Line et al., 1985).

of prepubertally castrated animals that displayed urine spraying and aggressive behavior in adulthood is the same as the percentage of animals that persist in these behavioral patterns after they have been castrated in adulthood (Hart and Barrett, 1973; Hart and Cooper, 1984).

These clinical surveys of horses and cats are along the same lines as experimental studies that have been conducted under controlled conditions. For example, in a study of the behavior of male dogs castrated at 40 days of age, but tested as adults, the prepubertal castrates mounted females and exhibited as much sexual excitement in the presence of females as intact control dogs (LeBoeuf, 1970). Bull calves castrated at four to five months of age were found to mount estrous females and exhibit pelvic thrusting sometimes as vigorously as intact bulls (Folman and Volcani, 1966). These observations, pointing out that prepubertal castration does not necessarily impair masculine behavior to any greater extent than does postpubertal castration, are paralleled by similar findings in rats and guinea pigs.

It appears that prepubertal castration is not qualitatively different from postpubertal castration, and the main influence, if any, of prepubertal castration is the time gained by castrating the animal early. That is, since castration often results in a gradual decline in various aspects of male behavior, castrating the animals prepubertally simply extends the time for the effect of castration to occur. Perhaps our concepts about the profound effects of prepubertal castration stem from their rather obvious effects on body size and shape of males, and we erroneously assume the same for behavior.

The effects of castration in adulthood are rather well documented with regard to sexual behavior because this behavior is so easily measured. In all species one or more aspects of mating activity decline after castration. During the period of decline, many animals will repeatedly mount and exhibit pelvic thrusting but are incapable of erection or ejaculation, whereas other males stop all sexual activity soon after castration. A few other individuals may engage in all mating responses, including the ejaculatory response, for months or even years after castration.

Aside from individual differences, there are pronounced species differences in the average persistence in sexual activity after castration in adulthood. This difference is illustrated in Figure 9-11. A striking example of the persistence of copulatory behavior following castration is evident in a study on goats. Of seven males castrated, six continued to copulate, on occasion, for over one year postoperatively (Hart and Jones, 1975).

Goats and other species that persist in copulatory activity do exhibit a decline in sexual motivation, as evidenced by a reduced frequency of copulatory attempts and usually an increased latency to initiate copulation when placed with a female (Figure 9-12). Some male dogs will continue to mate and to exhibit complete mating behavior for from two to five years following castration, but during this time there is a significant decline in copulatory frequency.

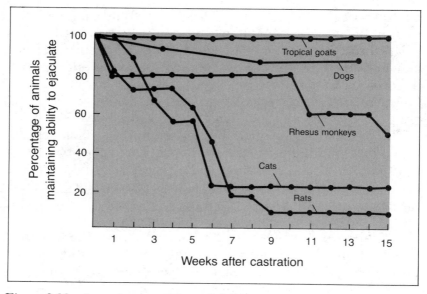

Figure 9-11 A species comparison of differences in persistence of copulatory behavior following castration. (From Hart, 1974, with data added from Hart and Jones, 1975.)

As will be discussed in the final section of this chapter, castration is an effective way of dealing with undesirable aggressive and urine marking behavior, as well as sexual behavior, in companion animals. The principles revealed by the study of castration effects on sexual behavior hold for all behavior that is affected by castration. For one thing, there is a high degree of individual variation within a species in the rate or degree of behavioral change following gonadectomy. For the most part, this individual variation cannot be related to the amount of prior sexual experience. There are also major species differences in the effects of castration.

■

BEHAVIORAL EFFECTS OF OVARIAN HORMONES

The ovarian hormones, estrogen and progesterone, are primarily of concern in the behavior of females. Estrogen is produced by the developing Graafian follicle and reaches its peak in secretion just before the onset of estrus. As far as is known, in all mammalian species (except the human female) an increase in estro-

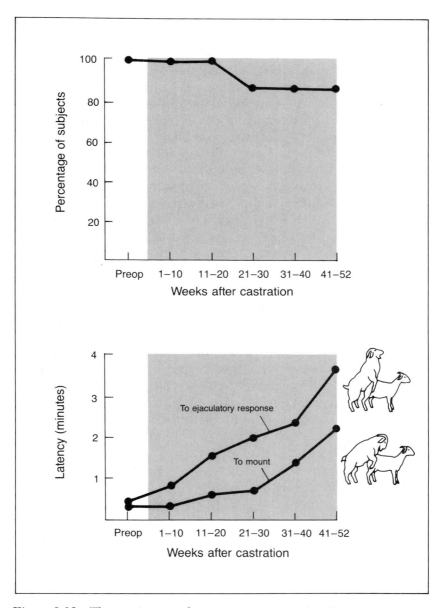

Figure 9-12 The persistence of mating activity in eight adult male goats following castration is evident in the top graph showing the percentage of males retaining the ability to copulate through postoperative testing extending for one year after castration. A reduction in motivation or sexual interest, however, is evident almost immediately after castration in the same male goats, as revealed by an increase in latency to initiate mounting, and to ejaculatory responses shown in the bottom graph. (From Hart and Jones, 1975.)

gen secretion is needed to bring on a full display of sexual behavior. In some species the secretion of progesterone seems to facilitate the display of behavioral estrus. The secretion of estrogen is controlled principally by the gonadotropins from the anterior pituitary; these hormones are in turn controlled by releasing factors from the hypothalamus. Temporal control of the estrus cycles in various species can be related to neural activity in the hypothalamus. Environmental influences, such as changes in day length, have influences on gonadotropin secretion through the hypothalamus.

Changes in sexual behavior brought about by the secretion of estrogen are those of increased attractiveness to males, initiation of proceptive responses such as seeking out males (if allowed to), and the display of receptive behavior in response to a male's sexual advances. Other changes not directly related to copulation are an increase in general activity and an increase in urination frequency (especially in the presence of the male) in some species.

Ovarian hormones, of course, influence behavior by acting on the brain. It has been shown that the implantation of solid estrogenic substances in the hypothalamus of spayed rats, cats, and rabbits will evoke receptive responses without any of the morphological signs of estrus in the genitalia (Barfield and Chen, 1977; Davis et al., 1979; Harris and Michael, 1964; Palka and Sawyer, 1966). The use of autoradiography has revealed the cellular uptake of radioactive hormones by individual hypothalamic neurons (Pfaff and Keiner, 1973).

Role of Ovarian Hormones in Sexual Behavior

The concept of sexual dimorphism and the role of androgen during the development of the brain have been dealt with in the discussion of male hormones and behavior. As far as is known, there is no behaviorally significant amount of ovarian hormone secretion during early life. This lack of secretion of any sex hormone leads to the development of a nervous system along feminine lines. In fact, the administration of fairly large doses of estrogen to neonatal female rats actually suppresses the full development of a feminine system.

Effects of Estrogen in Adult Females   In studies of the hormones necessary for the production of estrous behavior in spayed females, it has been found in all species that estrogen alone will induce receptivity if enough of the hormone is given. However, in rodents the amount of estrogen needed is much less, and the quality of the behavior better, if progesterone is also administered. For the artificial induction of estrus in rats, for example, a small amount of estrogen is given 48 to 72 hours before the desired time of estrus, and progesterone is administered to this "estrogen-primed animal" four to six hours before the desired period of estrus. Without progesterone, a larger dose of estrogen would be necessary to induce estrus. This type of hormonal treatment tends to mimic the blood concentrations of those hormones in naturally cycling females.

In sheep and goats, estrogen alone induces sexual receptivity, but prior treatment with progesterone for a week or two prevents the development of refractoriness to repeated treatments with estrogen (Robinson, 1959; Scaramuzzi et al., 1972).

Estrous behavior in cats and rabbits can be easily evoked without progesterone, and there are no reports of progesterone facilitation. In dogs, estrogen alone will evoke sexual receptivity, but an injection of progesterone the day before or the day of estrus seems to have a facilitatory effect with respect to the female's attractiveness to the male (Beach and Merari, 1970).

Just as estrogen can evoke sexual responsiveness in castrated males, so testosterone tends to mimic estrogen in evoking receptive behavior in the female. This has been documented in laboratory rodents and also in the ewe. Interestingly, in ewes treatment with a single dose of testosterone evokes normal female sexual behavior, but treatment daily for four weeks with testosterone results in the appearance of male sexual responses, including patterns of sniffing females, performing flehmen, and occasionally mounting. In the same study it was found that ewes treated with an estrogen daily for four weeks also displayed male behavior (Signoret, 1975). As will be explained later this phenomenon can be explained by the hormones first activating the neural basis of female behavior and subsequently activating the neural basis of male behavior.

Effects of Progesterone   With the exception of the special facilitatory effects on the induction of estrus in some species, progesterone has an inhibitory action on female sexual responsiveness, and it has some suppressive effects on male behavior as well. In laboratory rodents progesterone has been shown to inhibit male sexual, aggressive, and scent marking behavior. These effects occur apart from progesterone suppression of gonadotropin secretion. It is noteworthy, then, that progesterone has effects on behavior opposite those of both testosterone and estrogen. In large doses, progesterone produces general anesthesia in various animals, and there are indications that the hormone has calming or tranquilizing effects in somewhat lower doses.

Behavioral Effects of Ovariectomy

At the time of puberty the ovary of normal females starts secreting estrogen and progesterone in a pattern characteristic of the particular species. However, if the female is spayed during or after an estrous period, sexual behavior will not be displayed again. Females are thus considerably different from males, which continue to show sexual behavior for varying periods of time after gonadectomy. There are no reports of any enduring behavioral differences resulting from gonadectomy performed before or after the first estrus.

Experiments on rats have documented that ovariectomy leads to increased appetite, increased food intake, and decreased general activity. Theoretically, ovariectomy would have the same effect in other domestic animals, but experimental documentation of this is not available. The gain in body weight from enhanced food intake and reduction of general activity would logically be greatest in females that cycle throughout the year or for several months of the year, such as cows and horses. The effects of both ovariectomy and castration on body weight have been dealt with in Chapter 5.

Role of Ovarian Hormones in Maternal Behavior

Despite the long-standing success of endocrinologists in mimicking the hormonal changes that bring about male and female sexual behavior in many species, their success in producing completely normal parental or maternal behavior with exogenous hormones has been limited. The best example is the induction of maternal responsiveness in virgin female sheep with injections of estrogen and progesterone plus stimulation of the cervix with a vibratory stimulus (Keverne et al., 1983).

The only other species where there has been a concerted effort to artificially induce maternal behavior by hormone injections is the laboratory rat.

Maternal behavior in rats can actually be elicited on a non-hormonal basis in females that have never had a litter of pups by continuously exposing the females to infant pups for a period of up to six or seven days (Rosenblatt, 1967). Normally, however, females respond maternally to their offspring as soon as they are born, and it is believed that this immediate onset of postparturient maternal behavior is a reflection of a certain pattern of secretion of progesterone, estrogen, and prolactin. By injecting estrogen, progesterone, and prolactin in a schedule resembling that seen during gestation, maternal behavior can be evoked in ovariectomized female rats within a couple of days after being presented with rat pups (Moltz et al., 1970). This latency of just a day or two is still not as short as that existing after natural parturition, but it is shorter than the six to seven days required for nonhormonal induction of the behavior.

The shortest latency for the artificial induction of maternal behavior in rats was achieved in an experiment in which blood from female rats going through parturition was cross-transfused with blood in virgin females. In this experiment maternal behavior was evoked in a little less than 25 hours in the virgin females (Terkel and Rosenblatt, 1972).

■

APPLIED AND PRACTICAL CONSIDERATIONS

The opportunity to make the behavior of animals more suitable for their environment or to make it easier for us to control them by altering hormones is always a matter of great interest. Will castration cure a male dog of its aggressive behavior? Will spaying a female dog calm it? By the same token we wonder if abnormal hormone secretions, a so-called hormonal imbalance, can cause animals to act in unexpected ways. Almost all of these types of concern deal with the sex hormones. In this section we shall consider some behavioral problems that stem from abnormalities of sex hormone secretion and some practical concerns regarding castration and ovariectomy.

Perinatal Masculinization

The presence of androgen during the critical period of prenatal and postnatal life leads to the development of the predisposition for masculine behavior; the absence of androgen leads to the predisposition for feminine behavior. Abnormal behavior in females, characterized by an enhancement of masculine behavioral tendencies, may be brought about by disease conditions such as virilizing adrenal hyperplasia, in which the fetal adrenal cortex produces an excess of androgens. Masculinization may also occur in the developing fetus of domestic animals if a pregnant animal is given a substance with androgenic properties. Some progestational substances are structurally similar to androgens and if given to pregnant animals to reduce the chance of abortion can masculinize a female fetus.

Should Farm Animals Be Castrated?

In the husbandry of domestic animals, we should always be examining our practices. One routine practice that might be questioned is that of prepubertal castration of males intended for slaughter. A point often overlooked is that there are certain economic advantages to not castrating males. Bulls are more efficient at feed utilization and gain weight faster than steers (Champayne et al., 1969). Castration also involves extra labor and expense.

How feasible is it to keep herds of bull calves and yearling bulls? A study by Kilgour and Campin (1973) provides some information. They recorded fighting, other types of agonistic behavior, and mounting in a herd of two-and-one-half- to three-year-old Friesian and Jersey bulls in pasture and compared them with groups of older bulls. The young bulls maintained regular grazing cycles. There was little serious fighting and, in fact, the most frequent social encounters were classified as amicable. Aggressive activity and mounting increased in frequency and intensity as the bulls approached four years of age and eventually presented management problems.

Of course, young bulls that get along well on pasture may act differently in small corrals, and the frequency of injury to one another may be increased in feedlots. Since agonistic behavior increased with maturation, the age at which bulls must be marketed would be critical.

In an evaluation of the advantages and disadvantages of castrating farm livestock, Kiley (1976) points out some of the management procedures that would have to be examined in raising gonadally intact males for slaughter. Animal density would be an important consideration since excessive crowding increases aggressive encounters. Isolation of animals from a herd tends to increase the occurrence of aggression toward conspecifics, and so this would have to be watched. The way farmers handle and interact with their livestock influences their animals' reaction to them; thus precautions would be necessary to ensure the stockhandler always has a dominant position.

Behavioral Effects of Castration on Companion Animals

The behavioral effects of castration on sexual behavior serves as a useful model for a consideration of castration for certain types of problem behavior. Male cats are routinely castrated prepubertally. But often the operation is delayed until adulthood, when it is assumed that the castration will reduce roaming, urine spraying, and aggressive behavior. Similarly, adult male dogs are sometimes castrated in an attempt to reduce roaming, urine marking, aggressiveness, and mounting. The comparative effectiveness of castration in these two species is illustrated in Figure 9-13. Studies indicate that castration performed for problem behavior in cats is effective 80 percent to 90 percent of the time, whereas in dogs the operation is effective only about 60 percent of the time (Hart and Barrett, 1973; Hopkins et al., 1976).

Castration is also performed on stallions that are not particularly valuable breeding animals when sexual or aggressive behavior becomes difficult to manage. As in dogs and cats, the operation is effective much of the time but not 100 percent. Castration is effective in eliminating undesirable sexual behavior and aggression toward people in stallions that are displaying these problem behaviors between 60 and 70 percent of the time. The use of castration to eliminate objectionable aggression toward other horses is effective about 40 percent of the time (Line et al., 1984).

Horse people have often maintained that geldings which display stallion-like behavior have been castrated incorrectly, such that the epididymis (tubes conveying sperm from the testes to the spermatic cord) was not completely removed. It was mis-

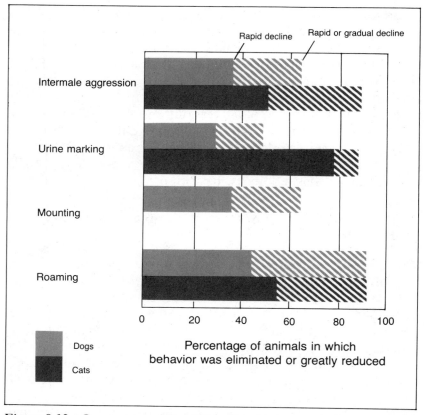

Figure 9-13 Comparison of the effects of castration on problem behavior in dogs and cats: When problem behavior was eliminated or greatly reduced, the rate of decline of the behavior was rapid in some cases and graded in others. The percentage of animals in which the decline was rapid or gradual is indicated by horizontal bars. (Graph from Hart, 1979, with data from Hart and Barrett, 1973, and Hopkins et al., 1976.)

takenly felt that the retained epididymis secreted testosterone and this caused the stallion-like behavior. The horseman's term for such an operation is "proud cut." The notion of proud cut horses having extra testosterone has been discounted by hormone assay procedures showing that geldings with stallion-like behavior and suspected of being proud cut had no more testosterone than regular geldings (Cox et al., 1973). The occurrence of stallion-like behavior in geldings is, of course, explained by noting that geldings do have the neural circuitry for male behavior as a function of perinatal androgenic effects. We have seen previously that the frequency of occurrence of male-like behavior in geldings is about the same as in stallions that are castrated postpubertally.

There is little evidence for alteration of behavioral patterns that are not sexually dimorphic. Therefore, generally desirable characteristics such as hunting ability and house guarding in dogs are not altered. Also, there is no evidence that older animals, with presumably more experience in performing the problem behavior, are affected any less frequently than younger, less experienced males. When behavior that we expect to be altered by castration actually persists, then it is not due to residual amounts of testosterone in the bloodstream. Testosterone is metabolized very rapidly and following castration, blood concentration of testosterone is reduced to nondetectable or behaviorally insignificant levels within hours. When objectionable sexually dimorphic behavioral patterns persist in castrates, the administration of synthetic progestins may sometimes suppress the behavior (Hart, 1980, 1981; Hart and Hart, 1985).

Behavioral Problems in Females Relating to Ovarian Hormones

In the discussion of ovarian hormones, mention was made of the behavioral influences of progesterone, especially the calming property. This is likely to be an important influence in bitches where high levels of progesterone continue for about 60 days after estrus. These levels of progesterone occur in non-mated bitches as well as in those that are pregnant (Smith and McDonald, 1974). If a female dog is spayed shortly after estrus, the sudden withdrawal of progesterone might be expected to bring about an emotional disturbance, with the possibility of irritability, aggression, and depression.

Experimental work on sheep showing that short-term estrogen treatment of females led to behavior typical of females in estrus but that long-term estrogen treatment brought on masculine behavior offers some insight into the abnormal behavior of females suffering from cystic ovaries. In cows with ovarian cysts, estrogen blood levels remain high for a long period of time (Glencross and Munro, 1974). The general term for the syndrome is nymphomania. Behavior associated with nymphomania begins with increased and prolonged sexual receptivity to males and ends with the display of bull-like behavior in long-standing cases. The bull-like behavior includes a deep voice, frequent attempts to mount other cows, and agonistic behavior such as pawing dirt. An explanation for the occurrence of first female behavior and then malelike behavior is that estrogen first extends the period of female sexual behavior, but the neural elements mediating female behavior become refractory to estrogen while neural elements that mediate male behavior become progressively more activated by the estrogen to the point where the females seem virilized.

REFERENCES

BARFIELD, R. J., AND J. CHEN. 1977. Activation of estrous behavior in ovariectomized rats by intracerebral implants of estradiol benzoate. *Endrocrinology* 101:1716–1725.

BEACH, F. A. 1970. Coital behavior in dogs. VI. Long-term effects of castration upon mating in the male. *J. Comp. Physiol. Psychol.* 70, (part 2):1–32.

BEACH, F. A. 1974. Effects of gonadal hormones on the urinary behavior in dogs. *Physiol. Behav.* 12:1005–1013.

BEACH, F. A., M. G. BUEHLER, AND I. F. DUNBAR. 1982. Competitive behavior in male, female, and pseudohermaphroditic female dogs. *J. Comp. Physiol. Psychol.* 96:855–874.

BEACH, F. A., AND H. FOWLER. 1959. Individual differences in the response of male rats to androgen. *J. Comp. Physiol. Psychol.* 52:50–52.

BEACH, F. A., AND R. E. KUEHN. 1970. Coital behavior in dogs. X. Effects of androgenic stimulation during development on feminine mating responses in females and males. *Horm. Behav.* 1:347–367.

BEACH, F. A., R. E. KUEHN, R. H. SPRAGUE, AND J. J. ANISKO. 1972. Coital behavior in dogs. XI. Effects of androgenic stimulation during development on masculine mating responses in females. *Horm. Behav.* 3:143–168.

BEACH, F. A., AND M. MERARI. 1970. Coital behavior in dogs. V. Effects of estrogen and progesterone on mating and other forms of social behavior in the bitch. *J. Comp. Physiol. Psychol.* 70, part 1:1–22.

BEATTY, W. W. 1979. Gonadal hormones and sex differences in nonreproductive behaviors in rodents: Organizational and activational influences. *Horm. Behav.* 12:112–163.

BEKOFF, M. 1974. Social play and play-soliciting by infant canids. *Amer. Zool.* 14:323–340.

BERNDTSON, W. E., B. W. PICKETT, AND T. M. NETT. 1974. Reproductive physiology of the stallion. IV. Seasonal changes in the testosterone concentration of peripheral plasma. *J. Reprod. Fert.* 39:115–118.

BLOCH, G. J., AND J. M. DAVIDSON. 1968. Effects of adrenalectomy and experience on postcastration sex behavior in the rat. *Physiol. Behav.* 3:461–465.

BREEDLOVE, S. M., AND A. P. ARNOLD. 1980. Hormone accumulation in a sexually dimorphic motor nucleus of the rat spinal cord. *Science* 210:564–566.

CHAMPAYNE, J. R., J. W. CARPENTER, J. F. HENTGES, A. Z. PALMER, AND M. KOGER. 1969. Feedlot performance and carcass characteristics of young bulls and steers castrated at four ages. *J. Anim. Sci.* 29:887–890.

CLARKE, I. J. 1977. The sexual behavior of prenatally androgenized ewes observed in the field. *J. Reprod. Fert.* 49:311–315.

CLARKE, I. J., R. J. SCARAMUZZI, AND R. V. SHORT. 1976. Effects of testosterone implants in pregnant ewes on their female offspring. *J. Embryol. Exp. Morph.* 36:87–99.

CLEMENS, L. G. 1974. The neurohormonal control of masculine sexual behavior. In *Reproductive behavior*, pp. 23–53, eds. W. Montagna and W. A. Sadler. New York: Plenum.

COX, J. E., J. H. WILLIAMS, P. H. ROWE, AND J. A. SMITH. 1973. Testosterone in normal, cryptorchid, and castrated male horses. *Equine Vet. J.* 5:i–iv.

DAMASSA, D. A., E. R. SMITH, B. TENNENT, and J. M. DAVIDSON. 1977. The relationship between circulating testosterone levels and male sexual behavior in rats. *Horm. Behav.* 8:275–286.

DAVIDSON, J. M. 1966. Activation of the male rat's sexual behavior by intracerebral implantation of androgen. *Endocrinology* 79:783–794.

DAVIS, P. G., B. J. MC EWEN, AND D. PFAFF. 1979. Localized behavioral effects of tritiated estradiol implants in the ventromedial hypothalamus of female rats. *Endocrinology* 104:898–903.

DE WIED, D. 1974. Pituitary–adrenal system hormones and behavior. In *The neurosciences, third study program*, eds. F. O. Schmitt and F. G. Worden. New York: Rockefeller University.

D'OCCHIO, M. J., AND D. E. BROOKS. 1982. Threshold of plasma testosterone required for normal mating activity in male sheep. *Horm. Behav.* 16:383–394.

DRORI, D., AND Y. FOLMAN. 1964. Effects of cohabitation on the reproductive system, kidneys and body composition of male rats. *J. Reprod. Fert.* 8:351–359.

FLETCHER, I. C., AND D. R. LINDSAY. 1971. Effect of rams on the duration of oestrous behaviour in ewes. *J. Reprod. Fert.* 25:253–259.

FLETCHER, T. J., AND R. V. SHORT. 1974. Restoration of libido in castrated red deer stag (*Cervus elephaus*) with oestradiol-17b. *Nature* 248:616–618.

FOLLEY, S. 1969. The milk-ejection reflex: A neuroendocrine theme in biology, myth and art. *Proc. Soc. Endocr.* 44:x–xx.

FOLMAN, Y., AND R. VOLCANI. 1966. Copulatory behaviour of the prepubertally castrated bull. *Anim. Behav.* 14:572–573.

GLENCROSS, R. G., AND I. B. MUNRO. 1974. Oestradiol and progesterone levels in plasma of a cow with ovarian cysts. *Vet. Rec.* 95:168–169.

GROSVENOR, C. E., AND F. MENA. 1967. Effect of auditory, olfactory, and optic stimuli upon milk ejection and suckling induced release of prolactin in lactating rats. *Endocrinology* 80:840–846.

GRUNT, J. A., AND W. C. YOUNG. 1952. Differential reactivity of individuals and the response of the male guinea pig to testosterone propionate. *Endocrinology* 51:237–248.

HARDING, C. F., AND H. H. FEDER. 1976. Relation between individual differences in sexual behavior and plasma testosterone levels in the guinea pig. *Endocrinology* 98:1198–1205.

HARRIS, G. W. AND R. P. MICHAEL. 1964. The activation of sexual behaviour by hypothalamic implants of oestrogen. *J. Physiol.* 171:275–301.

HART, B. L. 1974. Gonadal androgen and sociosexual behavior of male mammals. A comparative analysis. *Psychol. Bull.* 81:383–400.

HART, B. L. 1978. Reflexive mechanisms in copulatory behavior. In *Sex and behavior: Status and prospectus*, pp. 205–242, eds. T. E. McGill, D. A. Dewsbury, and B. D. Sachs. New York: Plenum.

HART, B. L. 1979. Problems with objectionable sociosexual behavior of dogs and cats: Therapeutic use of castration and progestins. *Comp. Cont. Educ.* 1:461–465.

HART, B. L. 1980. Objectionable urine spraying and urine marking behavior in cats: Evaluation of progestin treatment in gonadectomized males and females. *J. Amer. Vet. Med. Assoc.* 177:529–533.

HART, B. L. 1981. Progestin therapy for aggressive behavior in male dogs. *J. Amer. Vet. Med. Assoc.* 178:1070–1071.

HART, B. L., AND R. E. BARRETT. 1973. Effects of castration on fighting, roaming, and urine marking in adult male cats. *J. Amer. Vet. Med. Assoc.* 103:290–292.

HART, B. L., AND L. J. COOPER. 1984. Factors relating to fighting and urine spraying in prepubertally gonadectomized male and female cats. *J. Amer. Vet. Med. Assoc.* 184:1255–1258.

HART, B. L., AND L. A. HART. 1985. *Canine and feline behavioral therapy.* Philadelphia: Lea & Febiger.

HART, B. L., AND T. O. A. C. JONES. 1975. Effects of castration on sexual behavior of tropical male goats. *Horm. Behav.* 6:247–258.

HART, B. L., AND J. LADEWIG. 1979. Serum testosterone of neonatal male and female dogs. *Biol. Reprod.* 21:289–292.

HART, B. L., AND J. LADEWIG. 1980. Accelerated and enhanced testosterone secretion in juvenile male dogs following medial preoptic-anterior hypothalamic lesions. *Neuroendocrinology* 30:20–24.

HART, B. L., AND M. G. LEEDY. 1983. Female sexual responses in male cats facilitated by olfactory bulbectomy and medial preoptic/anterior hypothalamic lesions. *Behav. Neurosci.* 97:608–614.

HOPKINS, S. G., T. A. SCHUBERT, AND B. L. HART. 1976. Castration of adult male dogs: Effects on roaming, aggression, urine marking and mounting. *J. Amer. Vet. Med. Assoc.* 168:1108–1110.

IZARD, M. K., AND J. G. VANDENBERGH. 1982. The effects of bull urine on puberty and calving dates in crossbred heifers. *J. Anim. Sci.* 55:1160–1168.

JÖCHLE, W. 1975. Current research in coitus-induced ovulation. *J. Reprod. Fert. Suppl.* 22:165–207.

KATONGOLE, C. B., F. NAFTOLIN, AND R. V. SHORT. 1971. Relationship between blood levels of luteinizing hormone and testosterone in bulls, and the effects of sexual stimulation. *J. Endocrinol.* 50:457–466.

KEVERNE, E. B., F. LEVY, P. POINDRON, AND D. R. LINDSAY. 1983. Vaginal stimulation: An important determinant of maternal bonding in sheep. *Science* 219:81–83.

KILEY, M. 1976. A review of the advantages and disadvantages of castrating farm livestock with particular reference to behavioral effects. *Brit. Vet. J.* 132:323–331.

KILGOUR, R., AND D. N. CAMPIN. 1973. The behaviour of entire bulls of different ages at pasture. *Proc. New Zealand Soc. Anim. Prod.* 33:125–138.

LARSSON, K. 1966. Individual differences in reactivity to androgen in male rats. *Physiol. Behav.* 1:255–258.

LE BOEUF, B. J. 1970. Copulatory and aggressive behavior in the prepubertally castrated dog. *Horm. Behav.* 1:127–136.

LEE, V. W. K., I. A. CUMMING, D. M. DEKRESTER, J. K. FINDLAY, B. HUDSON, AND E. J. KEOGH. 1976. Regulation of gonadotropin secretion in rams from birth to sexual maturity. I. Plasma LH, FSH, and testosterone levels. *J. Reprod. Fert.* 46:1–6.

LEVINE, S. 1969. An endocrine theory of infantile stimulation. *Stimulation in early infancy*, pp. 45–63, ed. A. Ambrose. New York: Academic.

LINCOLN, G. A., AND W. DAVIDSON. 1977. The relationship between sexual and aggressive behavior, and pituitary and testicular activity during the seasonal sexual cycle of rams, and the influence of photoperiod. *J. Reprod. Fert.* 49:267–276.

LINDSAY, D. R., Y. COGNIE, J. PELLETIER, AND J. P. SIGNORET. 1975. Influence of the presence of rams on the timing of ovulation and discharge of LH in ewes. *Physiol. Behav.* 15:423–426.

LINE, S. W., B. L. HART, AND L. SANDERS. 1985. Prepubertal versus postpubertal castration of male horses: Effect on sexual and aggressive behavior. *J. Amer. Vet. Med. Assoc.* 186:249–251.

LIPTRAP, R. M., AND J. I. RAESIDE. 1978. A relationship between plasma concentrations of testosterone and corticosteroids during sexual and aggressive behaviour in the boar. *J. Endocr.* 76:75–85.

MASON, J. W. 1968a. Organization of the multiple endocrine responses to avoidance in the monkey. *Psychosom. Med.* 30:774–790.

MASON, J. W. 1968b. "Over-all" hormonal balance as a key to endocrine organization. *Psychosom. Med.* 30:791–808.

MAZUR, A., AND T. A. LAMB. 1980. Testosterone, status, and mood in human males. *Horm. Behav.* 14:236–246.

MOLTZ, H., M. LUBIN, AND M. NEUMAN. 1970. Hormonal induction of maternal behavior in the ovariectomized nulliparous rat. *Physiol. Behav.* 5:1373–1377.

MORFAN, P. D., G. W. ARNOLD, AND D. R. LINDSAY. 1972. A note on the mating behavior of ewes with various senses impaired. *J. Reprod. Fert.* 30:151–152.

PALKA, Y. A., AND C. H. SAWYER. 1966. The effects of hypothalamic implants of ovarian steroids on oestrus behavior on rabbits. *J. Physiol.* 185:251–269.

PARROTT, R. F. 1978. Courtship and copulation in prepubertally castrated male sheep (wethers) treated with 17β-estradiol, aromatizable androgens, or dihydrotestosterone. *Horm. Behav.* 11:20.

PFAFF, P. 1968. Autoradiographic localization of radioactivity in rat brain after injection of tritiated sex hormones. *Science* 161:1355–1356.

PFAFF, D., AND M. KEINER. 1973. Atlas of estradiol-concentrating cells in the central nervous system of the female rat. *J. Comp. Neurol.* 151:121–158.

ROBINSON, T. J. 1959. The estrous cycle of the ewe and doe. In *Reproduction in domestic animals*, vol. 1, pp. 291–333, eds. H. N. Cole and P. T. Cupps. New York: Academic.

ROSE, R. M., I. S. BERNSTEIN, AND T. P. GORDON. 1975. Consequences of social conflict on plasma testosterone levels in Rhesus monkeys. *Psychosom. Med.* 37:50–61.

ROSENBLATT, J. A. 1967. Nonhormonal basis of maternal behavior in the rat. *Science* 156:1512–1514.

SACHS, B. D., AND V. S. HARRIS. 1978. Sex difference and developmental changes in selected juvenile activities (play) of domestic lambs. *Anim. Behav.* 26:678–684.

SANFORD, L. M., W. M. PALMER, AND B. E. HOWLAND. 1977. Changes in profiles of serum LH, FSH and testosterone, and in mating performance and ejaculate volume in rams during ovine breeding season. *J. Anim. Sci.* 45:1382–1391.

SCARAMUZZI, R. J., D. R. LINDSAY, AND J. N. SHELTON. 1972. Effects of repeated oestrogen administration on oestrus behaviour in ovariectomized ewes. *J. Endocrinol.* 52:269–278.

SECCHIARI, P., F. MARTORAMA, S. PELLEGRIMI, and M. LUISI. 1976. Variation of plasma testosterone in developing Friesian bulls. *J. Anim. Sci.* 42:405–409.

SELYE, H. 1950. *Stress.* Montreal: ACTA.

SHARMA, O. P., AND R. L. HAYS. 1968. Release of oxytocin due to genital stimulation in bulls. *J. Dairy Sci.* 51:966.

SIGNORET, J. P. 1975. Effects of oestrogen and androgen on the sexual behaviour responses of the ovariectomized ewe. *Psychoneuroendocrinology* 1:179–184.

SMITH, E. R., D. A. DAMASSA, AND J. M. DAVIDSON. 1977. Plasma testosterone and sexual behavior following intracerebral implanation of testosterone propionate in the castrated male rat. *Horm. Behav.* 8:77–87.

SMITH, M. S., AND L. E. MC DONALD. 1974. Serum levels of luteinizing hormone and progesterone during the estrus cycle in pseudo-pregnancy and pregnancy in the dog. *Endocrinology* 94:404–412.

STERN, J. J., AND M. MURPHY. 1971. The effects of cyproterone acetate on the spontaneous activity and seminal vesicle weight of male rats. *J. Endocrinol.* 50:441–443.

STUMPF, W. E. 1971. Autoradiographic techniques for the localization of hormones and drugs at the cellular and subcellular level. *Acta Endocrinologica Suppl.* 153:205–222.

TERKEL, J., AND J. S. ROSENBLATT. 1972. Humoral factors underlying maternal behavior at parturition. Cross transfusion between freely moving rats. *J. Comp. Physiol. Psychol.* 80:365–371.

THOMAS, T., AND C. NEIMAN. 1968. Aspects of copulatory behavior prevent atrophy in male rats' reproductive system. *Endocrinology* 83:633–635.

VANDENBERGH, J. G. 1969a. Male odor accelerates female sexual maturation in mice. *Endocrinology* 84:658–660.

VANDENBERGH, J. G. 1969b. Endocrine coordination in monkeys: Male sexual responses to the female. *Physiol. Behav.* 4:261–264.

VANDENBERGH, J. G. 1977. Reproductive coordination in the Golden hamster: Female influences on the male. *Horm. Behav.* 9:264–275.

VOGLMAYR, J. K. 1975. Output of spermatozoa and fluid by the testis of the ram in response to oxytocin. *J. Reprod. Fert.* 43:119–122.

VOITH, V. L. 1979. Clinical animal behavior. *Calif. Vet.* 33(6):21–25.

VOM SAAL, F. S. 1981. Variation in phenotype due to random intra-uterine positioning of male and female fetuses in rodents. *J. Reprod. Fertil.* 62:633–650.

WARREN, R. P., AND L. R. ARONSON. 1956. Sexual behavior in castrated, adrenalectomized hamsters maintained on DOCA. *Endocrinology* 58:293–304.

WEISS, J. M., B. S. MC EWEN, M. T. SILVA, AND M. KALKUT. 1970. Pituitary-adrenal alterations and fear responding. *Amer. J. Physiol.* 218:864–868.

O ur domestic animals serve a number of indispensable functions in human society, and our history of living with them goes back to the earliest signs of civilization. By domesticating livestock mammals, early human societies were able to more predictably feed themselves and could do more work. Dogs were probably the first to be domesticated and assisted humans in herding other animals. Cats were among the last to be domesticated; they helped to reduce rodent pest problems. At present, some domesticated speices are still maintained to provide food and fiber products. Others serve as companions and may be treated as members of human families, a role in human society different from that initially associated with domestication.

Living with, caring for, and managing domestic animals demands that we be acutely concerned about their welfare. Of course, when the animals are of economic importance, such as livestock or valuable breeding pets, it would be inconsistent with economic policies not to look after their welfare in terms of feeding them, providing them with shelter, and keeping them free of diseases. Farmers pride themselves on providing food, shelter, and medical care sufficient to make the productivity of animals far exceed that which would be found in the wild. Our concerns in the area of animal welfare, however, must go beyond that of providing food, shelter, and freedom from disease to encompass preventing unnecessary suffering of animals from physical or emotional disturbances.

Before the domestication of animals, it would not have made sense for our ancestors to be concerned about animal welfare. Human beings evolved in an environment where wild animals were hunted because of nutritional needs or were fought off as predators.

With the advent of domestication we made our animals entirely dependent on us. Their bodies and their behavior have been genetically manipulated to fit the environments we have constructed (see Chapter 6). One of the hallmarks of artificial selection has been to reduce the fear that domestic mammals have of human beings, so that we can more easily approach and handle them. In wild animals this fear of people and other predators serves them well and is an important part of their strategy in staying alive. On the domestic scene we control an animal's food supply, its shelter, and even its health and freedom from disease. Animal welfare has therefore become an important aspect of management of domestic animals.

■

CONTROVERSIES IN ANIMAL WELFARE CONCERNS

Assuming that people who care for domestic animals have as a universal objective the desire to eliminate or reduce suffering of animals as much as possible, what we can glean from the study of animal behavior helps us to meet this objective. The purpose of this chapter is to present some background on how the study of animal behavior may guide our inquiries in this area. Recently the issue of animal welfare and animal behavior has become so emotional that there is an attempt to promulgate laws and federal guidelines, and to make public policy with very little background. With an understanding of the principles of animal behavior presented in this book, and an understanding of the legitimate controversies in animal welfare concerns that center around animal awareness, our right to kill (or not kill) animals, and animal care, we can promulgate laws and federal guidelines that are based on accurate facts and concepts so that we have what society wants and needs in the long run.

Animal Awareness

Those of us who have had pets or live closely with animals probably have no difficulty in accepting the notion that our domestic animals have some type of conscious awareness, and that when they have inadequate food or shelter or adverse behavioral experiences, they have an awareness of these experiences.

Recently people have increasingly felt that with regard to conscious awareness, humans are on a continuum with other animals. Our perception of the basic behavioral differences between humans and other mammals has narrowed. Some of the momentum was directly accelerated by Donald Griffin (1976), who exposed the close similarities between animal communication and human language and proposed that the differences in communication, like other behavioral differences, do not constitute a qualitative dichotomy. Providing numerous examples, Griffin has illustrated that animals do not respond mechanically to external (or internal) stimulation, without conscious understanding and intent, any more than humans do. He further suggested that awareness is biologically adaptive, since "understanding" the environment allows an animal to adjust its behavior accordingly. The interrelating of conscious experience with behavior would be reinforced by natural selection. Interestingly the acceptance of a human-animal continuum with regard to conscious awareness has probably promoted the field of sociobiology as spearheaded by Wilson (1975), whose main theme is that social behavior is subject to evolutionary processes and that animals, like humans, exhibit aspects of culture that respond to selection pressures.

There is not the space here to debate the intricacies of the awareness issue, of whether animals have the same kind of conscious awareness that human beings do or whether one species of domestic animals has a different kind of awareness than another species. The issues about awareness in animals involve some assumptions about mental states that are scientifically untestable, just as are assumptions about conscious awareness in human beings other than ourselves. For the purposes of this chapter, the working hypothesis shall be that animals are aware of their surroundings, aware of their feelings of hunger, physical discomfort, and disease state, and aware of unpleasant emotional states, all of which may be related to whether they are suffering or not.

Whether chickens have the same degree of conscious awareness as pigs, dogs, or cats is not a question that we can answer. One can make the case that, given differences in brain structure and what we know about the mediation of consciousness by the cerebral cortex as opposed to more primitive areas of the brain, there is a difference in conscious awareness between many of our domestic mammals and that of human beings. A critical point might be that our awareness allows us to be aware of our own suffering and to be aware of the awareness of suffering in others. One could argue that animals do not experience the latter. At any rate we shall assume that domestic mammals have some degree of conscious awareness and that when they suffer from inadequate food and shelter, poor health, or unpleasant emotional experiences they are aware of this suffering. Animal welfare deals with the issues of attempting to understand the basis of animal suffering, whether animals do in fact suffer under certain circumstances, and what might be done to alleviate the suffering that does occur.

Our Right to Kill Animals

Before we go into a discussion of the behavioral information bearing on animal welfare, we have to recognize that much of human emotion revolves around personal beliefs and value systems, which cannot be dealt with in a scientific way. Much of the discussion of animal rights is within this realm (Rollin, 1981). For example, some people object to the killing of animals for any reason. They believe that to take an animal's life is to violate a concern for its own rights to live and enjoy a full life span. Other people point to the fact that when an animal is killed swiftly, there can be no pain or suffering connected with the act of killing per se, and therefore the killing of animals does not involve the issue of suffering or animal welfare. Indeed, animals that are suffering from disease are frequently killed in a swift, and presumably painless, manner to reduce the suffering that they are experiencing. There are other people who believe that killing animals to alleviate long-term suffering is desirable, but that killing them to provide protein in the form of red meat is an infringement on their rights. Most people in the world would argue that an animal suffers no more when killed for the purposes of providing food than when killed to eliminate long-term suffering. Some livestock species are raised to supply human food, in the form of beef, lamb, pork, and poultry, and are killed before they reach sexual maturation. Dairy cattle, sheep kept for wool, and beasts of burden are maintained long into sexual maturity but are eventually killed in the same manner as those maintained for food production. Even the deaths of pets are often elected by their owners as a more humane alternative than allowing the pets to live in discomfort. If all the animals meet their death in the same way, can we say that one group suffers more than another?

The science of animal behavior cannot deal with value systems or personal beliefs involved in the issues just discussed above. We shall deal only with the issues of suffering while an animal is alive and assume that responsible husbandry provides a swift and painless death when the animal's time has arrived. Of course, no life, not even that of the most sheltered human beings, is without some degree of physical or emotional suffering. We can hardly expect animals to live a life that is completely free of any kind of suffering any more than we can expect this of people. But our goal with animals is to keep suffering within reasonable limits, as we would with our fellow human beings.

Double Standards

One of the emotional issues that comes up with regard to animal welfare is the favored status that some animals receive. The fact that we regard some mammals as food products, others as family members, and still others as pests (rats, mice) confounds and complicates our efforts to develop clear criteria for appropriate welfare practices. Assuming that the common mammals have about the same degree of awareness of suffering, our concern necessarily encompasses the livestock species as well as the household pets that share our homes. Pet dogs and cats currently enjoy much more sympathy with regard to their freedom from suffering than do livestock animals and a number of wild species. With very little public concern, rats are systematically poisoned in cities and farms with materials that can produce a rather prolonged illness that finally leads to the death of the animal. Despite the fact that one could argue that the awareness of a rat is about equal to that of a dog or cat, if dogs and cats were killed with rat poison the way that rats are, there would be a considerable public outcry about the prolonged suffering from internal hemorrhages produced by the anticoagulants that constitute the most frequently employed form of rat poison.

Young animals generally arouse more sympathy than do older ones in the minds of many people. Yet it makes good sense to attribute a higher level of awareness to animals that have lived longer and experienced more than young animals.

Another double standard reflects the fact that rare and beautiful animals usually engender stronger campaigns for their freedom from suffering than do plentiful and less beautiful animals. Laws in almost every country protect endangered species from being excessively harassed or hunted. However, any suffering that might be brought about by hunting is experienced by the more plentiful animals just as readily as it is by the animals on our endangered-species lists.

Clearly we see all around us the application of double standards in the formulation of public policy and laws stemming from personal beliefs and value system judgments rather than from principles derived from studies of animal behavior. The most a behaviorist can do is to point out that there is a double standard involved and possibly to encourage leaders in society to adopt uniform policies in animal welfare legislation.

Animal Care

Since people have cared for domesticated mammals throughout history, we are constantly faced with choices regarding the quality of care we provide. Both with our livestock species and our pets we select and determine their living space, diet, and social contacts. Since there are conventional procedures for animal care, we usually just adopt these conventions.

In this century procedures for animal production have undergone rapid technological development, with a parallel accelerated change in typical animal care methods. At the same time the size of the animal husbandry industry has expanded with the growth in population and meat consumption. In comparison with the field of livestock production, behavioral science is new and has not kept fully abreast of the behavioral effects of various methods of caring for animals. Nevertheless, behavioral science can now contribute to our understanding of the effects of the rapid growth and change in the animal husbandry industry.

The practical reality is that animal husbandry involves a huge number of animals that must each be housed, fed, cleaned, shorn, or milked, whatever is appropriate, and eventually butchered or otherwise disposed of at death. Economic considerations of course play a role in selecting husbandry practices, but, whatever procedures are used, the task is a complex logistical accomplishment.

A number of questions merit exploration by behavioral scientists. Some of these are highlighted below.

Thermoregulation Freely moving animals in varied environments are able to walk over and stand by a tree to avoid the hot sun or strong wind or stand in a puddle to cool off. In these ways they can behaviorally adjust environmental variables such as temperature, humidity, sunlight, and wind. As animals behaviorally express their preferences in these ways, the adjustments can often be shown to result in reduced mortality or improved health for the animals. Some experiments illustrate this point. Pigs in one study used a combination of techniques to thermoregulate, remaining near outside heaters in the day and huddling in a hut at night (Ingram et al., 1975). Lactating cows given access to shade in a subtropical summer were found to have a higher conception rate and a lower incidence of clinical mastitis, as well as lower body temperatures compared with cows not given access to shade (Roman-Ponce et al., 1977). Although one might expect that the behavioral regulation of temperature by animals would always be adaptive, unshorn adult sheep do not always use available shelter to protect themselves from severe weather. During a storm, recently shorn ewes in one study used high grass when it was available as shelter and suffered lower lamb mortality than unshorn ewes (Lynch and Alexander, 1976).

Housing Space and Restraint One consequence of reducing the size of animal pens is that the animal no longer has the option of behaviorally regulating its temperature. Further, the cage may be so small or the animals so numerous that an animal is effectively physically restrained. High density per se does not necessarily imply negative effects on animals (Freedman, 1979). For example, young zebu bulls housed together in fenced paddocks in small groups later mated with expertise and panache, but other bulls grazed in a herd on the range exhibited timidity when they first met a cow in estrus (MacFarlane, 1974). We might guess that crowding would adversely affect some species but have a negligible effect on others, especially those with a natural herding tendency.

Movement   From time to time it is necessary to transport or regroup animals. This shifting of animals to unfamiliar settings or groups stresses them and often is accompanied by additional stresses such as thermoregulation and restraint. The growth of piglets is adversely affected by changing their quarters whether or not strange pigs are introduced (Charlick et al., 1968; Dantzer, 1970). Transporting of pigs occasionally even precipitates sudden death (Fraser et al., 1975). Transporting has been shown to increase the incidence of *Salmonella* excretion (Williams et al., 1970) in pigs.

Recreational Uses   In contrast with the other uses of animals, in sporting events entertainment is the primary issue, rather than the economics of production of meat or other animal products. In a bullfight, a dog fight, or the use of bulls, calves, and horses at a rodeo, the sport brings up the importance of humans' challenge to dominate beasts more than 10 times their weight or strength (Lawrence, 1982). Some people are concerned about whether welfare issues are involved in these events.

In the sections below we shall deal with three perspectives in the area of animal welfare. The first is the practice of making comparisons or analogies with ourselves regarding animal suffering. The second perspective involves a consideration of the relation between animal disease, ill health, and the physiology of suffering. Third, we shall look at what the behavior of the animal tells us about its suffering and the advantages and disadvantages of using behavioral signs as possible markers of suffering. A very systematic treatise on these and other perspectives with regard to interpreting experimental work on animal suffering has been written by Marian Stamp Dawkins (1980), and this book is recommended to those who want to read more about the subject. Some of her observations and arguments have been drawn upon in the comments in this chapter.

■

COMPARISON OF ANIMALS WITH OURSELVES

Probably the most frequently used approach to determine whether animals are suffering or not is to make a comparison of their status with that of ourselves. In a sense we cannot escape this approach. When we see a cat or a horse display movements or postures that we ourselves would display when painfully stimulated, and when we see the painful stimulus, we cannot but help assume that the animal is suffering as we would. This example is rather simple and straightforward and would probably be an assumption that a behaviorist would be comfortable with. Other comparisons, however, become much more subjective. When we walk into a barn in which the ambient temperature is 10°C (50°F), we are naturally cold and could easily say that were we to live in the barn 24 hours a day, we would obviously be suffering. To assume that cattle living in this environment are suffering because of the low temperature is another matter. Their metabolic rates are different, and the insulation they derive from skin and fur and, in general, their thermal control mechanisms probably allow them to be quite comfortable in this environment. To say that "If I lived in such a cold barn I would be suffering, therefore the cows must be suffering," is clearly an erroneous assumption.

Are there any scientifically legitimate reasons to believe that we can say anything about an animal's experience in suffering by comparing the animal with ourselves? In favor of this approach are some striking similarities between certain parts of the human brain and those of the common domestic mammals. The brain structures that lie beneath the cerebral cortex, commonly referred to as brain stem structures, are morphologically quite similar in both humans and domestic animals. It is in these more primitive subcortical areas of the brain that the transmission and mediation of pain responses occur. Furthermore, some of the basic emotional responses involved with fear and escape have their neural circuitry within the subcortical structures. Therefore, in dealing with stimuli that we relate to the production of pain or fear we feel somewhat justified in assuming that animals are experiencing some of the same emotions that we would.

Species-Typical Behavior and Suffering

The difficulty with trying to conclude whether an animal is suffering by comparison with ourselves occurs in areas where there are not clear behavioral markers of pain. To assume an animal is experiencing suffering as a result of confinement, isolation, certain environmental temperatures, or lack of activity (boredom) because we would under the same circumstances is to be naive about all of the innate differences between humans and domestic mammals. Animals have evolved a variety of species-typical behavioral patterns that suit them for certain natural environments. We must assume that whether they experience emotional stress or not relates to their species-specific adaptive behaviors. Thus, highly social animals such as dogs or horses may suffer, one could argue, if deprived of contact with members of their own species (or substitute members such as humans) for a prolonged period of time. One would not expect an asocial species, like the cat, to suffer as an adult if deprived of contact with members of its own species. Therefore, meaningful understanding of the relation between potential suffering and the animal's environment must take into account its own native behavioral life style and the suitability of its behavioral patterns for this environment.

We must also understand a good deal about animal behavior to interpret vocalizations and facial postures before we prematurely assume that certain behavioral signs are evidence of underlying suffering. What sounds like a cry to us, simply because it resembles a cry from a baby, may represent a simple exchange of signals between two animals. Puppies may yelp when simply being picked up because this represents an alarm signal that attracts the mother and in the wild state would have some advantage to the young. One could hardly argue that picking up a puppy is causing it to suffer. On the other hand, yelping may be displayed when an animal is suffering from the wounds sustained in an automobile accident.

Even evidence of physical injury does not necessarily imply that the animal has suffered. For example, we know that people may be wounded while playing at a sport and not notice the wound until there is a break in the activity. On the other hand, if one carefully observed the wound being inflicted this would be painful and the individual would be suffering.

Cross-Species Use of Analogies

Our domestic animals differ from one another in brain and behavioral function, but not as much as we differ from them. Before we can assume that we can use ourselves as models for the possibility of animals' suffering, we might try to imagine whether we can use a cat as a model for suffering in a horse or a pig as a model for a cow. If we had (miraculously) a very clear idea of the awareness of the potential suffering going on in the mind of a pig, but not in that of a horse, could we use a readout of the mental events in a pig to tell us when a horse is suffering? This illustrates the conceptual difficulties of using the behavior of one species, namely ourselves, in figuring out the degree of suffering in other species.

■

ANIMAL HEALTH AND SUFFERING

Disease or physical injury is recognized as one of the main causes of suffering in humans. It is quite legitimate that we should also consider that these factors cause suffering in animals. We can argue that it is even adaptive for an animal to suffer from illness. If diseases did not cause depression and loss of appetite, animals would not reduce their activity, a necessary step in recovery from disease. The pain and general malaise that accompanies some diseases and contributes to an animal's suffering are necessary for an animal's survival, just as the experience of pain is necessary for an animal to keep weight off a wounded foot.

Since gross disturbances in health or the presence of injuries can reasonably be said to cause suffering in animals, one of our responsibilities in the area of animal welfare is to eliminate and reduce the amount of diseases and injuries to which animals may be subjected. However, there is an objection to using only outright disease or injury as markers of suffering, which is that they are probably too insensitive as measures of long-term suffering (Dawkins, 1980). If animals are in an environment that may eventually lead to a diseased state, must we wait until they actually come down with a disease before we say they have been suffering? Are there any preclinical physiological signs of illness that might be used as markers for animal suffering?

The growing productivity of farming is sometimes presented as evidence that animals are healthy, receive excellent care, and are not suffering. However, in the modern sense, productivity refers to production per unit of cost, and productivity could simply reflect that the animals were not allowed to exercise, resulting in a rapid weight gain. Measures of productivity may also be heightened by reducing overhead costs, such as animal housing and farm labor. When large numbers of animals are involved, various conditions could result in an erosion of health for a substantial proportion of the animals without impacting on the immediate productivity figures. Theoretically, conditions associated with high productivity could eventually be seen to lead to a diseased state were it not that animals were sold off for market sooner. Therefore, productivity alone is not necessarily evidence of optimal husbandry conditions. The reasons behind the productivity must be considered before assuming that animal welfare is satisfactory (Dawkins, 1980).

Measures of Stress

The most frequently studied physiological process that can be related to suffering is stress. We know that prolonged behavioral and physical stress may impair the body's immunological system and in other ways lead to sufficient breakdown of an animal's resistence that the animal becomes diseased. One might say then that an environment that produces stress capable of eventually producing a disease state might be causing suffering prior to the disease.

The most widely accepted sign of stress is an abnormally high level of adrenal corticosteroid output (see Chapter 9). This is why in so many experiments this is used as a critical measure of stress. The measurement of stress by corticosteroid production gives us a means of working backward in time to link disease with suffering. For example, certain housing or shipping conditions can be related to a certain level of stress as indicated by analysis of corticosteroid output. If past experiences or experiments have also shown that this level of stress usually leads to serious sickness or death, we could then say that the conditions produced some suffering and that we should look into altering the situation.

Some experiments serve to illustrate how environmental situations can affect coticosteroid levels. For example, corticosteroid levels are increased in sheep that are removed from their flock or transported by truck and in calves that are transported

(Kilgour and deLangen, 1970; Reid and Mills, 1962; Crookshank et al., 1979). A simple alteration in routine management can induce increased plasma cortisol levels in heifers of several breeds (Adeyemo and Heath, 1982). Likewise, in calves a variety of normal husbandry practices are capable of inducing changes in heart rate and plasma levels of corticosteroids (Stephens and Toner, 1975). A number of these conditions or procedures also may lead to disease, reduced productivity, or increased mortality.

It would be useful to develop and agree on some standardized measures of the stress experienced by an animal, such that comparisons within and across species could easily be made. Perhaps certain of the domestic species are especially vulnerable to any disruption, and others are more resilient. In sheep, certain individuals readily become extremely agitated, and there is an increase in heart rate response. Some sheep respond with low-level excitement, but show the same elevated heart rate. Only completely unexcitable sheep fail to show the heart rate response (Syme and Elphick, 1982/1983).

We must be very careful in interpreting what parts of the environment are stressful to an animal. We might think, for example, that if sheep are chased by a dog, loaded into a truck, or dipped into an insecticide vat that this would be more stressful than the wool-shearing process. One study revealed, however, that the shearing process, which involved separating the sheep from the rest of the flock, produced the highest level of adrenal corticosteroid output when compared with the other procedures (Kilgour and deLangen, 1970).

The approach of measuring corticosteroid output allows us to ask the animal if it is physiologically distressed, and to what extent. This objective quantitative measure, although somewhat inconvenient to obtain, can provide direct information concerning various animal care methods, such as whether a specific housing is stressful or whether a certain bedding material or type of heating mitigates the stress.

Apart from heart rate, which can be measured nonintrusively, a variety of blood parameters have been measured in studies of stress effects. Chronic stress has also been experimentally associated in pigs with increased plasma levels of total protein and glucose and decreased urea (Barnett et al., 1982/83). It would be useful to identify the measures that are the most accurate across species in yielding a finely tuned assay of an animal's stress level.

Stress That Is Beneficial

Milder levels of stress that have no long-term health consequences would not be considered to reflect major suffering. We must recognize that animals sometimes seek a degree of excitement that leads to some moderate stress. Humans themselves will seek stressful situations to escape from continuous boredom. Therefore, it is reasonable to assume that a moderate degree of stress from time to time is beneficial; this is the "happy medium principle" (Fraser et al., 1975).

■

ANIMAL BEHAVIOR AND SUFFERING

To what degree can we use a animal's behavior as an indication of its welfare and whether or not it is suffering? Are there certain behavioral responses that indicate an animal is suffering from an unpleasant emotional state? In many ways, using an animal's own behavior to indicate its freedom from emotional distress is more reliable than attempting to evaluate its welfare by comparing its reactions to ours or even trying to establish a link between its physiological responses and its welfare.

There are several ways of looking at an animal's behavior and judging its welfare. We might compare the behavior of domestic animals in the captive state with that of their wild ancestors, who lived free of human intervention. Second, we might look for abnormal behavioral patterns that reflect disturbed emotional states. Third, we might test animals on their choice of environments or circumstances as a way of deciding if they were suffering in one of the environments or circumstances.

Behavior of Animals in Captivity versus That of Their Wild Counterparts

Since domestic animals in captivity have much less freedom and experience less varied environments than those in the wild, some people feel that the animals must be suffering since they are prevented from displaying and experiencing the entire range of their behavioral repertoires. There is no doubt that animals kept in all-male groups will not have the opportunity to experience sexual behavior or social interactions with members of the opposite sex (see Chapter 3). When we remove the young from mothers a day or two after birth, as is common on dairy farms, the females never have an opportunity to display maternal behavior (see Chapter 4). Also, most captive animals do not exercise as much as those in the wild. Do animals suffer from the lack of opportunity to engage in these more natural or instinctive urges?

As Dawkins (1980) has mentioned, there is undoubtedly something very attractive about the idea of wild animals roaming freely in their natural environment, able to feed at will and to mate and bear their young without interference. Can we really say, however, that providing hay to a horse while preventing it from foraging on its own leads to suffering? Does lack of opportunity to mate in a male animal that is castrated (and probably has very little sexual interest) lead to suffering? On the domestic scene we prevent predation of ungulates by carnivores and thereby spare the ungulates the fate of being captured and killed in the way that their wild ancestors were. One can argue that death at the hands of a predator constitutes a prolonged and painful experience, before the animal is rendered unconscious. In this respect the domestic environment is more humane than the wild one. We prevent the killing of songbirds by domestic cats in the name of curtailing the cruelty perpetrated by this type of predatory behavior.

But the case has been made in some instances that domestic animals are so severely deprived of an opportunity to display native behavior patterns that they do indeed suffer. For example, chickens that are kept in small battery cages cannot move more than a few steps nor flap their wings. They cannot dust bathe and they have nowhere to roost. Sows that are kept in farrowing crates may not even be able to turn around.

To what extent might these practices be considered inhumane? One could argue that the behavior of domestic animals is so altered genetically from that of the wild counterparts that a comparison with animals living in the wild is unreasonable. Modern-day chickens, for example, will not sit on a clutch of eggs as did their wild jungle fowl ancestors. This lack of broodiness is a reflection of intentional genetic alteration through domestication. One would be hard pressed to say that chickens suffer from a lack of opportunity to incubate their eggs. Most domestic animals could not survive in the wild if turned loose. This is a reflection of both behavioral and morphological alteration through the process of domestication.

To what extent are wild animals free from suffering? Obviously wild animals are continuously exposed to the hazards of disease and predation. Often our captive animals are healthier and better off than wild ones because ailments may be recognized and treated expediently whereas wild animals may be left to a lingering death. Eventual death of most wild animals in fact probably results from capture by a predator or from suffering due to disease. We should not fall into the trap of thinking that a natural life is better simply because it appears to be more romantic to us as outsiders (Dawkins, 1980).

One might think of sheep as leading a romantic life, wandering freely over hillsides and unconstrained by cages or other attributes of modern farming. It has been pointed out, however, that mountain sheep may suffer more in the winter than animals kept under intensive husbandry systems (Ewbank, 1974). They may be almost eaten alive by blowfly maggots in the summer and suffer from cold and starvation in the winter (Tudge, 1973). There is too much suffering in wild animals for comparison between them and captive ones to be used as a standard of animal welfare without other supporting evidence.

Historically, the scientific comparison of wild and domestic situations has been retarded by the separation between the animal scientists and veterinarians, who study improving domesticated animal production and health, and the ethologists and psychologists, who tend to study the behavior of undomesticated species (Kilgour, 1976). Further, it is important to know not only the natural behavior of the animal but also the various consequences of the behavior, particularly those that increase fitness. Then, in creating environmental designs for domesticated animals, while understanding behavioral consequences, productivity, fitness, and welfare can be an integral part of the design (Baxter, 1982/1983).

It is tempting to make comparisons between wild and domestic animals in looking at restricted locomotion of animals and in evaluating housing conditions. However, because of the existence of genetic differences between wild and domestic animals or of differences between the wild environment and the domestic one, we cannot use wild animals and the wild environment as the main point of reference for whether or not domesticated animals are suffering. The evidence for suffering must rest on a number of parameters, among which can be the comparison of wild and domestic situations.

Display of Abnormal Behavior

Abnormal behavior might be defined as a behavioral pattern shown by a minority of the animal population that is not due to any obvious disease state, that is not adaptive in an evolutionary sense, and that is not maintained by a concurrent conditioning reinforcer or caused by identifiable events in early experience.

Let us look at a type of abnormal behavior that is obviously related to suffering. Under some circumstances animals have been known to bite into their own flesh and create such massive wounds that they die from their own self-mutilation. Most of us would not have much trouble stating that with this type of abnormal behavior the animal is experiencing suffering. Animals may also mutilate each other to the point of causing the death of a member of their own species. Under intensive housing, for example, pigs will sometimes bit each others' tails. Once the tail biting of certain pigs has reached a critical point, other pigs join in and the animal, which may be severely wounded, may die. Again, most of us would feel that this behavior leads to suffering by the victims. In this case we can point to housing conditions with concrete floors that preclude the animals from engaging in normal rooting or chewing behavior and state that the ensuing boredom leads to the abnormal biting behavior. It is known that giving pigs some "playthings" to chew on or soil to root in greatly reduces the behavior (Van Putten, 1969).

How often is abnormal behavior related to suffering? The term itself is quite loaded, and sometimes people automatically associate the occurrence of abnormal behavior with the experience of suffering. A couple of examples serve to point out that abnormal behavior is not always related to suffering. Flank sucking in Doberman pinscher dogs has been described as abnormal behavior in that it occurs in only a minority animals of one particular breed, and the behavior occurs in a variety of environments including when the dogs' owners are around or not (Hart, 1977). Flank sucking appears not to be harmful to the dogs in that the skin is usually not irritated and the animal shows no outward sign of pain. The behavior is not caused by boredom or lack of exercise, because even some of the most active Dobermans still engage in the behavior.

Chewing wool is another abnormal behavior pattern that occurs in some Siamese cats. Like the flank-sucking syndrome in Dobermans, it occurs in only one breed (Hart, 1976a). As long as large amounts of wool are not consumed, the behavior produces no signs of suffering by the cats, although it may occasion much suffering in owners, whose wool stockings and sweaters are ruined.

Another type of abnormal behavior that has been mentioned in the animal welfare literature includes the stereotypes that are common in caged animals, particularly those in zoos. The most common stereotyped movement is pacing backward and forward along the side of the cage. This behavior appears to reflect the boredom or frustration of a captive life and we are led to wonder whether such animals are suffering. Again, by general consensus, we could probably say that if the behavior were continued to such a degree that the animal wore down its paws until they were bleeding, it would be suffering. Short of this, can we actually evaluate the degree of frustration or boredom that the animal is undergoing? Can we say that a moderate degree of stereotyped pacing is related to suffering any more than flank sucking in dogs?

Another problem in using abnormal behavior as a guide to detecting the occurrence of suffering is that one cannot always tell if the behavior is even abnormal. For example, a wide variety of attention-getting behaviors have been described in dogs, ranging from chasing shadows, biting in the air, and lameness to medical signs such as coughing or diarrhea (Hart, 1979). Even bizarre, abnormal-looking behavioral patterns may be adopted as a means of getting extra attention, sometimes in the form of comfort and affection, and at other times in the form of being yelled at. There is one reported instance of self-mutilation in a cat, which chewed at its tail to get attention from its owner (Hart, 1976b).

Perhaps it would be reasonable to at least accept a conservative approach for determining if abnormal behavior is related to suffering. This would involve again using physical health as a guide and extending it backward in time to include the occurrence of abnormal behavior. Thus, if the abnormal behavior would lead to serious illness, injury, or death if continued, then the abnormal behavior itself could be taken as reflecting an underlying state of suffering. Measuring physiological parameters indicating stress, such as adrenal corticosteroid production, would also help in diagnosing whether the abnormal behavior was endangering an animal's health. Beyond this we can only infer that if an animal is engaging in abnormal behavior, it could be because of boredom or frustration, but it becomes a guess as to whether the animal is suffering or not.

Animal Choice Experiments

The most direct approach of determining whether an animal is suffering or not would be comparable to asking our fellow humans if they were suffering. Such verbal communication is, of course, impossible. But as we have seen with experiments in discrimination learning, animals can be trained to reveal their choices by voting with their feet as it were (see Chapter 8). If we give a pig the choice between a concrete floor and a dirt floor, we can note its choice and draw conclusions as to which type of floor it would rather be on. Such a choice could reveal that it was suffering on the concrete floor, as it might if it were receiving electric shocks there. On the other hand, its choice of dirt might reflect a rather mild preference for dirt over concrete. If it had no particular preference, it would probably spend about an equal amount of time on each surface.

The Role of Choice Experiments Animal choice experiments are about the best thing we have to gain insight into an animal's subjective feelings. Some investigators have done experiments in which pigs were allowed to regulate their environmental temperature by manipulating a device that controlled a heat lamp or a ventilation fan. This is like asking an animal to indicate the temperature it prefers. Various other experiments along this line have examined an animal's choice of wide-open spaces versus confinement, the amount of daylight it receives, and the type of foodstuffs it prefers.

We usually feel that an animal chooses an environment that will promote its survival and optimal reproduction. In other words, an animal's choice is guided by inherited behavioral predispositions not only to avoid suffering but also to choose the best of two favorable environments. This behavior has paid off in terms of reproductive output. Animal choice experiments do not tell us if an animal is suffering in one environment or if it simply prefers one over another. The beauty of choice experiments is that we need make no assumptions as to what is rewarding or pleasant or unpleasant to animals. Nor do we need to put ourselves into the animal's shoes and attempt to figure out what it is experiencing by comparison or analogy to our own feelings.

But this is why animal choice experiments are fraught with problems when it comes to relying on them to tell us about animal suffering. Let us examine these problems further. Consider the fact that most pet dogs strongly prefer T-bone steak to kibble (coarsely ground meal or grain) dog food and would make this choice in laboratory test. The test results would probably be reinforced by behavioral signs in that in going for steak the dog would probably wag its tail, drool, and jump around. Does this mean that the dog is suffering when we feed it kibble dog food? Certainly not. Could we say that a dog was suffering if it never tasted steak for the rest of its life? Again, the answer would be no. However, if the kibble food was always laced with quinine or cayenne peppers and was very aversive, and in fact was leading to malnutrition and eventual premature death, then the experiment with steak versus kibble food would yield results consistent with the notion that the animal was suffering with the kibble diet. Obviously, choice experiments cannot be used as the sole basis for examining an animal's experience of suffering.

Animal choice experiments have been used to correct some of our own misconceptions about an animal's welfare. A rather well-known example dealt with a federal committee in England that had strongly recommended that the floors of battery cages for chickens not be made of small-gauge hexagonal wire but should be made of heavy, rectangular metal mesh because the committee felt the heavy mesh looked more comfortable. The committee-style floor, had it been widely adopted, would have been extremely expensive, particularly because it had a tendency to crack eggs. What bothered some animal behaviorists was the lack of evidence that the hens preferred to stand on the committee's floor rather than on the small-gauge wire. In a subsequent experiment hens were given an opportunity to choose between both types of floor, and they clearly preferred to stand on the small-gauge wire (Hughes and Black, 1973). Furthermore, photographs taken from under the floor showed that the hens standing on the smaller mesh received more support for their feet than those standing on the floor recommended by the committee. Although this experiment had a bearing on a legal ruling, it did not address the question of whether the hens were suffering on either one of the types of flooring. The floor choice experiment simply provided the basis for discarding more expensive flooring on the basis that the animals did not prefer the more expensive floor to begin with.

Experiential Effects and Animal Choice It is almost impossible to design a animal choice experiment without bringing into the picture an animal's past experience. Consider another experiment with chickens that allowed hens a choice between a battery cage and an outside run in the garden (Dawkins, 1977). Up to the time of the experiment, the hens had been raised in either battery cages or outdoor runs. The choice was presented to them in the form of a corridor in which by looking in one direction the hens saw the battery cages, and in the other direction the outside garden. The hens had an overall tendency to choose the outside run, but the choices were strongly influenced by their prior living environment. Hens that had been living in battery cages tended to choose the battery cage environment, whereas those raised in the outside environment chose the outside run. As the experiment was continued, and the hens accumulated experience in the two environments (each animal was confined for five minutes in whichever environment it chose), the hens raised in the battery cages came to choose the outside run. Again, this experiment tells us nothing about the strength of the preference, nor whether the animals were suffering in the battery cages. It does illustrate how an animal's previous experience enters into its choice.

Short-Term versus Long-Term Preferences in Animal Choice
We must not lose sight of the fact that animals change in their preferences over time. An animal that has become sexually exhausted would not choose to go into an environment with females as it would have before sexual activity had begun. Juvenile males that had not experienced a pubertal rise in testosterone would often choose different stimuli or environments than gonadally intact males. Similarly, in a two-choice test, a spayed female cat that chooses to avoid sexually active males is not suffering from a lack of sexual activity.

Animals Do Not Always Choose What Is Best   Herbivores sometimes choose to eat poisonous plants. Cattle are prone to bloat
if left to graze on green alfalfa for too long. We have seen in a
number of experiments that animals do not always select the
foods that promote the best health. This phenomenon is especially noteworthy in the human species. We can recognize
evolutionary adaptiveness for the existence of behavioral patterns that paid off in survival value or reproductive success in
the wild but that are of no value on the domestic scene. Saccharin, which has a sweet flavor and for which pigs have a strong
liking, has no food value. But in the natural environment sweet-
tasting substances are almost always associated with beneficial
carbohydrate foods that animals need for energy. This is an example of an isolated stimulus to which the animal responds from
a genetic background linking it to its wild ancestors, but that is
counterproductive on the domestic scene. There are other examples. We must occasionally subject animals to painful medical
experiences for their own health and welfare that, if the animals
were left to their own devices, they would clearly choose to
avoid. Consider the dipping of cattle to eliminate external parasites, injecting of vaccines to establish immunity, and restraining animals for treating wounds.

■

CONCLUSIONS

When our own species had fully domesticated several other species to make our own lives more enjoyable, we induced genetic
changes in the animals and controlled their environment to the
point where they became almost completely dependent on us
(see Chapter 1). We find ourselves now with the clear respon-
sibility of looking after their physical and emotional welfare. To
go to the extreme of granting animals the same rights as we have
would be counter to the very reasons for domesticating them.
And, if we do not domesticate animals, we are back to the pre-
dator-prey relationship, which pretty much excludes welfare
concerns. Animal welfare as it relates to the occurrence of a
concern for suffering is an aspect of the domestication phenomenon.

Presumably only the human species has an awareness or the
potential for awareness in animals of suffering. By the same

token we must be aware of the logical approaches to establishing guidelines for the identification of animal suffering and of ways to ameliorate or eliminate suffering. Although this is an area where scientific experiments are of limited value because of difficulties in interpreting results, using experiments and the wealth of behavioral information available is the best basis for sound progress in this field. We have examined three approaches that are used to evaluate animal suffering. These are (1) comparisons with ourselves, (2) animal health and suffering, and (3) animal behavior and suffering. No approach alone can yield solid conclusions. Therefore, a combination of approaches is necessary.

A logical starting point for diagnosing conditions that cause suffering uses, as a point of reference, conditions that would eventually lead to disease, injury, or death. This criterion offers testable hypotheses that can be evaluated by unbiased investigators. The degree of animal suffering that should be tolerated, or allowed to be offset by economic losses, pertains to the area of value systems and personal beliefs and is beyond the scope of behavioral science.

There are two goals that can be met by behavioral scientists. The first goal is to aid in the development of uniform standards of minimum care for all species, assuming equal awareness and potential for suffering so that livestock have a basic baseline of protection as well as pets. The second is to lead the way in establishing standards for diagnosing suffering as a step toward assessing specific complex situations as to their appropriateness for the species in question.

References

ADEYEMO, O., AND E. HEATH. 1982. Social behavior and adrenal cortical activity in heifers. *Appl. Anim. Ethol.* 8:99–108.

BARNETT, J. L., P. H. HEMSWORTH, AND A. M. HAND. 1982/1983. Effects of chronic stress on some blood parameters in the pig. *Appl. Anim. Ethol.* 9:273–277.

BAXTER, M. R. 1982/1983. Ethology in environmental design for animal production. *Appl. Anim. Ethol.* 9:207–220.

CHARLICK, R. H., H. R. LIVINGSTON, A. MC NAIR, AND D. W. B. SAINSBURY. 1968. The housing of pigs in one pen from birth to slaughter. *Exp. Farm Buildings II.* Sisloe, Australia: Nat. Inst. Agric. Eng.

CROOKSHANK, H. R., M. H. ELISSALDE, R. G. WHITE, D. C. CLANTON, AND H. E. SMALLEY. 1979. Effect of transportation and handling of calves upon blood serum composition. *J. Anim. Sci.* 48:430–435.

DANTZER, R. 1970. Retissement du comportement social sur le gain des poids chez des porcs en croissance. II Perturbation liees au melange d'animaux et au changement de loge. *Ann. Rech. Vet.* 1:117.

DAWKINS, M. 1977. Do hens suffer in battery cages? Environmental preference and welfare. *Anim. Behav.* 25:1034–1046.

DAWKINS, M. S. 1980. *Animal suffering: The science of animal welfare.* London: Chapman & Hall.

EWBANK, R. 1974. Clinical signs of stress in farm animals—changes in behaviour. *Brit. Vet. J.* 130:90.

FRASER, D., J. S. D. RITCHIE, AND A. F. FRASER. 1975. The term "stress" in a veterinary context. *Brit. Vet. J.* 131:653–662.

FREEDMAN, J. L. 1979. Reconciling apparent differences between the responses of humans and other animals to crowding. *Psychol. Rev.* 86:80–85.

GRIFFIN, D. R. 1976. *The question of animal awareness.* New York: Rockefeller University Press.

HART, B. L. 1976*a*. Behavioral aspects of selecting a new cat. *Feline Pract.* 6(5):8–14.

HART, B. L. 1976*b*. Social interaction between cats and their owners. *Feline Pract.* 6(1):6–8.

HART, B. L. 1977. Three disturbing behavioral disorders in dogs: Idiopathic viciousness, hyperkinesis and flank sucking. *Canine Pract.* 4(6):10–14.

HART, B. L. 1979. Attention-getting behavior. *Canine Pract.* 6(3):10–16.

HUGHES, B. O., AND A. J. BLACK. 1973. The preference of domestic hens for different types of battery cage floor. *Brit. Poult. Sci.* 14:615.

INGRAM, D. L., D. E. WALTERS, AND K. F. LEGGE. 1975. Variations in behavioral thermoregulation in the young pig over 24 hr periods. *Physiol. Behav.* 14:689–695.

KILGOUR, R., AND H. DELANGEN. 1970. Stress in sheep resulting from management practices. *Proc. New Zealand Soc. Anim. Prod.* 30:65–76.

KILGOUR, R. 1976. The contributions of psychology to a knowledge of farm animal behaviour. *Appl. Anim. Ethol.* 2:197–205.

LAWRENCE, E. A. 1982. *Rodeo: An anthropologist looks at the wild and the tame.* Knoxville: University of Tennessee Press.

LYNCH, J. J., AND G. ALEXANDER. 1976. The effect of gramineous windbreaks on behaviour and lamb mortality among shorn and unshorn Merino sheep during lambing. *Appl. Anim. Ethol.* 2:305–325.

MAC FARLANE, J. S. 1974. The effect of two post-weaning management systems on the social and sexual behaviour of Zebu bulls. *Appl. Anim. Ethol.* 1:31–34.

REID, R. L., AND S. C. MILLS. 1962. Studies on the carbohydrate metabolism of sheep. XIV. The adrenal response to psychological stress. *Aust. J. Agric. Res.* 13(1):282–295.

ROLLIN, B. E. 1981. *Animal rights and human morality.* Buffalo, N.Y.: Prometheus Books.

ROMAN-PONCE, H. H., W. W. THATCHER, D. E. BUFFINGTON, C. J. WILCOX, AND H. H. VAN HORN. 1977. Physiological and production responses of dairy cattle to a shade structure in a subtropical environment. *J. Dairy Sci.* 60:424–430.

STEPHENS, D. B., AND J. N. TONER. 1975. Husbandry influences on some physiological parameters of emotional responses in calves. *Appl. Anim. Ethol.* 1:233–243.

SYME, L. A., AND G. R. ELPHICK. 1982/1983. Heart-rate and the behaviour of sheep in yards. *Appl. Anim. Ethol.* 9:31–35.

TUDGE, C. 1973. Farmers in loco parentis. *New Scient.* 60:179.

VAN PUTTEN, G. 1969. An investigation into tail-biting among fattening pigs. *Brit. Vet. J.* 125:511–517.

WILLIAMS, L. P., JR., AND K. W. NEWELL. 1970. *Salmonella* excretion in joy riding pigs. *Am. J. Public Health,* 60:96–99.

WILSON, E. O. 1975. *Sociobiology: The new synthesis.* Cambridge, Mass.: Harvard University Press.

NAME INDEX

SUBJECT INDEX

reproductive seasonality, 85
selective breeding, 215–216, 229
semen collection, 119
sexual activity, 85
sexual behavior, female, 87,
    103, *104*, 105–106, *107*
sexual dimorphism, 310–311
sexual dysfunction, 125
sexual stimuli, 95
sperm depletion, 119
testosterone concentration, *319*
social behavior
    aggression, idiopathic, 35
    aggression, predator-prey, 32
    aggression, territorial-social, 34
    attention-getting behavior, 362
    breed behavior differences,
        232–236, *235*
    fighting, 26
    grooming, 57
    ground scratching, 40–50
    growling, 22
    human dominance over, 71
    olfactory ability, 43
    piloerection, 20, 22
    play, 57–58, *58*, 60
    scent marking, 44, *45*, 47–48,
        49
    scooting behavior, 51
    social organization, 63, 70, 120–
        121
    submissive behavior, 24, *25*
    territoriality, 61
    threat, 20
    visual displays, 39
    vocalization, 38–39
wild ancestors, 7–8, 86
Domestication
    behavioral aspects, 228
    history, 1, 345–346
    stages of, 5–6
    theories of, 3–6
Donkeys, domesticated, 3

Ejaculation, 116–119
Elephants
    domesticated, 3
    female grouping, 62
Estrous cycles, 96–97, *97*

Feral animals and domestic
    animals, 3
Feral asses and harem social
    grouping, 63

Feral cats, 3
    maternal behavior, 136
    solitary-territorial grouping, 61
Feral dogs, 3
Feral goats, 3, 152
Feral horses, 3
    dunging areas, 205
    harem social grouping, 62–63
    leadership hierarchies, 64
    placentophagy, 140
    scent marking, 48
Feral sheep, 3
    maternal behavior, 136
Ferrets
    domesticated, 3
    reflex ovulators, 308
Fish
    goldfish and operant learning, 289
    minnows and pheromones, 46
Flehmen, 46, 47, 90–94, 311 (*see
    also* Cats, Cattle, Dogs,
    Goats, Sheep, Swine)
Foxes, and death feign, 24
Foxes, red
    burying prey, 181
    male sex pheromones, 89
Freemartin heifer, 323–324
Fulani herdsmen, 18, 56, 65, 66,
    147, 148, *149*, 184, 186

Gazelles and fighting behavior, 27
Geese and maternal imprinting,
    252–254
Goats
    eliminative behavior, 205
    feeding behavior
        rumination, 186–188
        selective foragers, 185
        specific hungers, 190–193
        subordinates' diets, 178–179
    feral, 3, 152
    learning behavior
        early attachments, 252, 259
        operant learning, 291
    maternal behavior
        abnormal fetus presentation,
            139
        kid vocalizations, 143
        milk ejection, 148
        maternal labelling of young,
            146
        mother-young bonding, 146
        nursing-suckling relationship,
            150–151